Man's Place

Intelligence in Whales, Dolphins, and Humans

MAN'S PLACE

Intelligence in Whales, Dolphins, and Humans

by Karl-Erik Fichtelius and Sverre Sjölander

Translated from the Swedish by
Thomas Teal

LONDON
VICTOR GOLLANCZ LTD
1973

Originally published in Sweden as *Människan, kaskelotvalen och kunskapens träd* by Wahlström & Widstrand.

ISBN 0 575 01550 0

Printed in Great Britain by
Lowe & Brydone (Printers) Ltd., Thetford, Norfolk

When Gulliver had told his host in the land of the Houyhnhnms something about life and people at home in England, the latter observed that "he looked upon us as a sort of animals to whose share, by what accident he could not conjecture, some small pittance of reason had fallen, whereof we made no other use than by its assistance to aggravate our natural corruptions, and to acquire new ones which nature had not given us; that we disarmed our selves of the few abilities she had bestowed; had been very successful in multiplying our original wants, and seemed to spend our whole lives in vain endeavors to supply them by our own inventions."

Foreword

The Portuguese man-of-war is a kind of jellyfish consisting of a large floating air-sack to which a great number of different types of individual animals are attached. These individuals have various specialized functions: locomotion, reproduction, nutrition, or protection. Every man-of-war is a floating colony of specialized individuals who are physically united with each other and who carry on an intimate cooperation by means of a system of stimulus and response. The problem of coordinating the individuals within this entity appears to be a fairly simple one.

It is not necessary that the animals be physically united with one another in order for them to accomplish this kind of coordination, including the division of labor. When the individuals are separated from each other we no longer refer to the group as an entity but call it a society or community.

The most striking examples of animal societies are to be found among the insects—among termites, ants, and bees. Here the individuals are physically independent, but still very limited in their behavior. They are governed by communications systems that, in principle, resemble the ones governing the Portuguese man-of-war: stimulus-response systems of various levels of complexity. No degree of individual freedom worth mentioning exists within these insect societies.

There is undeniably a greater degree of individual freedom in a human society than in an insect society. A soldier ant remains a soldier ant all its life. A human soldier often stays in uniform, of course, but he can break loose and change his profession. In the more prosperous parts of the world he can also look forward to a pension. But for the vast majority of people on earth, the choices are extremely limited. And, in principle, human society's means of communication and control are the same as those of the ant community and the Portuguese man-of-war—stimulus and response.

In one respect, the ant society has an advantage over modern human society! The ant community has developed over the course of millions of years, and the individual ants are well adapted to it. Our human anthills have come into existence in the last quivering second of evolution. And individual adjustment is consistent with that fact, even though man, by virtue of his intelligence and his hands, is maybe the most adaptable of all the animals. The modern city is not a harmonious expression of mankind's biological heritage, perhaps no more than the thousand-stall barn is an expression of the temperament of a cow.

The great majority of people are irritated by this kind of comparison. They consider it quite unreasonable to talk about ant society and human society in the same breath, and they cite all manner of circumstances pointing to our utterly superior control of different situations and our extensive freedom of choice. The circumstances they cite are correct, today, for certain wealthy human societies and certain social classes.

But they are not true of other societies, other social classes, and other times. If we look at human society as a super-organism, the freedom of choice will appear to be considerably more modest than what obtains at the individual level. How else can we explain the fact that two-thirds of the earth's population is at present starving, or that in India, every year, 14,000 children go blind from undernourishment? If human society had freedom of choice to the extent we like to believe, then of course we would never have permitted such a state of things to come about.

We human beings will probably never work out the serious problems of coexistence until we recognize our limitations, both on the individual and the group level. This book about man and the sperm whale is written against the background of the difficult situation in which we find ourselves. By attacking the deeply rooted notion that man is the most remarkable thing God ever created, it should help us to understand that we are a part of biology. The book is propagandistic, but not in the sense that it compromises what the authors think is true or scientifically probable. There are highly speculative passages here and there. We hope it will be entirely clear to the reader when we are speculating and when we are providing what we consider to be facts. There are also anecdotal passages here and there. In so far as possible, these anecdotes have been chosen critically, and as a rule the source has been given. It is our hope that in the near future science will be able to show us which of these anecdotes we can believe and which we must regard as unbelievable.

The reader of this book should realize that the authors do not pretend to deal with all the problems of man's existence. We are quite conscious of the fact that we can only supply material on certain biological aspects of these problems, and that economic, historical, sociological, linguistic, and many other aspects can have an equal or greater importance. All of man's acquired knowledge must be brought into the discussion. The reader should also understand that when we call attention to the similarities between human beings and animals in some certain respect, thesc similarities apply only in that one respect and cannot be carried over into other areas. If we maintain that human beings and baboons have certain behavioral characteristics in common, then of course that assertion does not apply to certain other behavioral characteristics. One must not read between the lines in a book of this kind. Like most similar books, this one is not only the work of its authors but also of all those people with whom the authors discussed the book as it was being written. Moreover, several people have helped us directly with their critical comments on parts of the manuscript. We prefer not to mention these people by name, so as not to make them responsible for what we have produced. They saw only parts of the manuscript, and some of their suggestions were ignored.

Mr. Åke Dintler of the University Library in Uppsala collaborated on the chapter about the sperm whale's history. It was his thankless task to hunt out cold facts that might serve as a basis for a meaningful discussion of the whale's own history. As will be seen from Chapter 6, the results were meager. It would be unjust in the highest degree to blame Åke Dintler for that, and I owe him my profound thanks for his devoted work.

Karl-Erik Fichtelius
Uppsala, February 1971

Contents

PART I

The Crown of Creation–
Man or the Sperm Whale

CHAPTER ONE

Man's Place on Earth

The Biological Book of Genesis

By way of introduction, let us think of the earth as a living entity, the largest form of life we know. Lewis Thomas, an American biologist, sees its birth, briefly, as follows.

First the gigantic boulder cooled and water collected over a large part of its surface. Under the influence of solar radiation, inorganic material somewhere in the water was transformed into the kinds of compounds necessary for the beginning of life. It is not clear whether nucleic acids, the bearers of genetic information, appeared first and gave rise to proteins, or whether the proteins came first and produced nucleic acids. Whichever the case, the compound molecules called polymers could now proliferate. The earth's liquid layer became an active broth of biochemical reagents still without organized structure—a kind of primitive organism. It's hard to think of this enormous fluid mass as a living thing, but it's equally hard not to.

The polymers became membranes, and enzymes were joined to these membranes in various ways. Cells were produced, and reproduction began. The first cells were rudimentary and probably all of a similar type. A single successful cell may have been the ancestor of everything that was to come, but it may also have been the case that different ancestral cells developed in many different warm spots in the seas.

Whether there was one or many primordial cells, they must all have begun their existence in darkness, or near-darkness, because at the surface of the water the sun's rays were deadly. The atmosphere was much too thin to give protection. Respiration had not yet begun. The amount of oxygen in the atmosphere must have been less than one one-thousandth of what it is today.

The earth began as an organism that lived without oxygen, without breath. Before it could begin to breathe, oxygen had to be produced, and that required green plant cells. They turned up about 2.7 billion years ago. With their appearance, at a depth of maybe thirty feet in shallow places, the atmosphere began to change. When the oxygen concentration had increased enough to permit the formation of cells that breathed, the groundwork for a more rapid evolution had been laid. Cell aggregates formed that could live at the surface of the oceans, protected from the sun's radiation by the oxygen-rich atmosphere.

Respiration, evolution, movement—life existed everywhere. The forms were different, but the general arrangements for propagation, metabolism, growth, and mutation remained remarkably similar. Specialized forms flourished and then died out, always in some way recycled back into life. In the seas, which teemed with life, a steady cloud of dead plants and animals sank down through the depths all over the earth and formed deposits thousands of meters thick, which in turn became a source of new life.

Soil became what it is today—that cold, damp material you can get your hands into—full of living organisms but also of dead substances in transit, waiting to be used again. This is the planet's protoplasm.

The earth's respiration seems to be a cyclical phenomenon, like our own breathing. First the oxygen content reached the level we denote as critical for respiratory life, approximately one percent. This permitted an explosive expansion of life in the seas. When the oxygen content had risen to 10 percent, plants and animals moved onto dry land, and life spread quickly. At one point there was more oxygen in the air than there is today, about 20 percent. Then something happened. Carbon dioxide was consumed by the dominant plants, and the amount of carbon dioxide in the atmosphere decreased. By obstructing loss of heat, carbon dioxide has a hothouse effect on the earth. This hothouse effect was now reduced, the earth cooled down, and an ice age began. As a result, there was a reduction in plant photosynthesis, and the oxygen content of the atmosphere went down. The earth lost its breath, so to speak. It was on such an occasion that the giant reptiles disappeared from sight. Some scientists claim that rhythmic changes of this kind have taken place ever since life began. These fluctuations are the result of life itself, and every breath the earth takes produces enormous changes in evolution.

That's how the American biologist tells the story.

We must think along new lines about our place on earth. In the first place, we do not own it. We are not its masters. It was not created for us to play with, or consume, or vandalize. We are a part of it. We came out of it, and we ought to be proud and humble as we trace ourselves back to the first polymers in the first seas. The earth evolves, and in this stage of evolution, *Homo sapiens* has come to dominate almost all other life. But we are ourselves caught up in the earth's evolutionary process. A fresh old truth begins to emerge all the more clearly from the scientific research of the last decades—the understanding that no matter what form of life we study, from the virus to the redwood, we are actually studying ourselves.

The Population Explosion—Our Basic Environmental Problem

If rats in a limited area have an unlimited supply of food and water, their numbers increase at first relatively slowly, then more and more rapidly. Finally the growth curve levels off and the size of the community becomes constant. Crowding has at this point become so great that the animals cease to function optimally in various areas. One of these areas is reproduction.

The wolf pack in the Chicago Zoo has quite an ample but nevertheless limited space at its disposal. It has unlimited access to food and drink. But over the years the animals have not increased in number the same way rats do under similar conditions. The wolves appear to be capable of some kind of birth control—the dominant female decides which of her sisters will bear litters in any given year, and the dominant male controls his subordinates. It looks as if the wolves are capable of stopping the growth curve before it stops itself for lack of space. Whether these mechanisms are hereditary or to some extent learned is, under the circumstances, of secondary importance.

Mankind's situation cannot be compared either to the rats or to the wolves. Not all people today have unlimited supplies of food and water, but we are all aware of the fact that the number of human beings on earth is increasing explosively. Everyone realizes that we have to check the growth of our population before it reaches a level corresponding to the maximum number of rats. But politicians and different kinds of scientific specialists sometimes give the impression that it is their goal to bring mankind to just such a level of overcrowding. Every attempt to increase the earth's yield must be accompanied by serious efforts toward a more equable distribution of what does exist, and toward a defusing of the population bomb.

The central environmental problem of the future is the number of human beings on earth. If we become too numerous we will be forced to devastate the earth's resources. We will have to concentrate on certain useful plants and eradicate most of the others. We will have to destroy "useless" animal species and preserve certain others that, from our point of view, are effective sources of protein. In our zeal, we will disturb the biological balance to the point that we will in all probability meet with enormous catastrophes. When, later on, we require a particular grass or beetle or other living things as the point of departure for some new biological improvement, these creatures will long since have been wiped out. In the long run, no environmental policy will do any good unless it is coupled with an effective program of birth control.

Cultural Evolution Is Much Faster Than Biological Evolution

The increase in the earth's population is partly a function of time and of man's singular cultural development. This cultural evolution has been enormously more rapid than its sluggish biological counterpart. While it took about a million years for man to evolve from the higher primates, we have covered the distance from the Stone Age to the Atomic Age in less than ten thousand. We can compare the development of architecture with the progressive structural changes in the evolution of animal skeletons. The changes from Greek and Roman temples to Gothic and Late Gothic cathedrals are very great compared with the structural changes in the skeleton. And yet they occurred in the course of some twenty centuries, while the skeletal changes from fish to mammal took more than two hundred million years.

Man's hands have been both his blessing and his ruin. Without our hands, we would not have been able to bring about the cultural evolution we have accomplished. The arts of writing and printing, essential prerequisites for that evolution, are both based on the function of the hand—strictly speaking, on the development and function of a single muscle *(musculus opponens pollicis)* which moves the thumb against the other fingers.

The dangers such a rapid development pose for mankind have been the subject of lively debate ever since the Second World War. I will illustrate what I mean with one example, perhaps the most serious threat of all—the combination of human devotion and developing weaponry.

Human Devotion Pro and Con

In his book *The Ghost in the Machine,* Arthur Koestler has a chapter entitled "The Pathology of Devotion," in which he maintains that people's inclination to cooperate with one another is infinitely more dangerous than their selfishness. That may sound like a paradox, but historians are agreed that only a small percentage of historical violence has been the result of selfishly aggressive impulses. Wars have always been fought primarily for the gods, for king and country, for mankind's future happiness, or something of that kind. The number of people who have fallen victim to thieves, gangsters, rapists, and other criminals has always been negligible in comparison to the masses slain with a clean conscience in the name of the true religion, a just politics, or the right ideology.

To the seven deadly sins Koestler adds an eighth, the deadliest of them all: misplaced devotion. A young American sociologist, Lionel Tiger, gives biological arguments for supposing that this eighth deadly sin is more pronounced in men than in women. He may be right. But it might be wise to avoid the hot potato of sexual differences in this connection. I will content myself with trying to demonstrate that man's devotion, both men's and women's, has deep biological roots.

Modern biology has shown quite clearly that function, or behavior, is a leading factor in evolution. It has also shown that natural selection applies primarily to groups of individuals and not to individuals themselves. There are countless examples of how the individual's interests must give way to the interests of the group or of the species.

Among the primates there are examples of many different types of social organization. The lovable chimpanzees seem to base their society on their own kindhearted disposition. There is a certain order of dominance among the individuals in a chimpanzee group, but it is not strict. The meeting of two groups stirs up a tremendous emotional storm which, however, does not seem to include any antagonism, and the encounter often ends with everyone eating together in the same tree. Chimpanzees exhibit a glimmer of the primate potential for that built-in amiability we associate with the lost paradise—which does not inspire any great hope for the chimpanzee's survival. Despite its intelligence and strength, the chimpanzee is close to extinction. The capacity of the chimpanzee group for concerted action is not impressive. If an individual senses danger, it hides before giving a cry of warning—if it gives a cry of warning at all. The savannah chimpanzee usually just leaves its companions to their fate. This friendly, in many ways admirable, and second most intelligent of the primates is an evolutionary failure. Either it never had an ability to be socially effective, or it has lost it.

Baboons are a different matter. Baboon society, an extremely strict and consistent oligarchy, is an evolutionary bull's-eye, and baboons are found all over Africa. The baboon group is a match for the leopard and, over the years, it has waged and continues to wage a successful war against man. The key to success lies partly in the fantastic solidarity that has been chiseled out over millions of years. A single baboon is a dead baboon.

What's true of the baboon is also true of man. In order to survive the battle against an unfriendly environment, solidarity was a requirement. A single human being was a dead human being. Darwin's contemporary Alfred Russel Wallace, unfortunately forgotten for about one hundred years, wrote in 1864:

> In proportion as physical characteristics become of less importance, mental and moral qualities will have increasing importance to the well-being of the race. Capacity for acting in concert, for protection of food and shelter; sympathy, which leads all in turn to assist each other; the sense of right, which checks depradation upon our fellows . . . are all qualities that from earliest appearance must have been for the benefit of each community, and would therefore have become objects of natural selection.

Wallace's thoughts made no impression in 1864. But modern ethology has established that he was right. It is almost undoubtedly true that qualities such as courage and the capacity for friendship, commitment, and loyalty have been cultivated in human society. The tribe having the most members with these qualities won the battle.

We have a great problem in the fact that, biologically, we are equipped for life in relatively small groups but live in super-tribes. For 10 to 20 million years, our ancestors hunted together in small groups, and devotion to this little group has been written into our germ plasm. Much of what happens in politics can be traced back to this fact. In the super-tribes we comprise today, our leaders must constantly fight against the tendency toward disunion. Specialized pseudo-tribes, such as social groups, professional groups, and sporting groups, crystallize within the super-tribe. This re-creates tribal identity and permits such groups to go on living happily within the super-tribe. But more serious splintering also occurs. Empires split into independent states. Better means of communication and common goals are no help. The tendency to disunity is there, and sooner or later it will tip the scales. The largest states require pressure from outside in order to survive. They seem to need war, or its equivalent.

Swift cultural evolution has now turned devotion into a threat to man's existence. If the people in one of the superpowers commit themselves to the idea that the people in the second or third superpower must be exterminated and to that end display courage, self-sacrifice, and loyalty, and if at the same time the people in the enemy camp show tremendous solidarity in their defense as well as in their desperate counterattack, then I am afraid there will be very little left of what we have built. There is a serious risk that the surviving baboons would defeat the human beings still alive in the ashes. It has happened before that a dominant species has vanished from the face of the earth. People who are faithless and disloyal, those who were formerly unsatisfactory, would then have the greatest chance of surviving and saving our culture.

The situation is grave even without atomic weapons. The time is probably past when the wealthy nations could, with force of arms, hold back the third world, the hungry world. Modern biological weapons make it possible for small countries to make devastating attacks on the superpowers.

We will probably never be able to solve our problems in a satisfactory way and preserve our respect for our fellow men unless we reconsider our place on earth. We must be made to realize that we do not own the earth, but that we live here and in all probability will have to live here forever.

CHAPTER TWO

Large-Brained Animals

The Comparative Anatomy of the Cerebral Cortex

We are all pretty much agreed that the higher vertebrates exhibit feelings. Pigs, alligators, rats, and birds all look, at least, as if they became just as angry or frightened or satisfied as we do. But the "higher" mental processes are a different matter. We consider human beings to be more intelligent than chimpanzees, and chimpanzees in turn to be brighter than dogs and cats, at least when measured by our standards. The conclusion to be drawn from this is obvious: those parts of the brain that account for higher mental activities are, in all likelihood, the ones that show the greatest increase in volume and weight from the least intelligent animals up to man, while those parts of the brain that are engaged in sense reactions are probably the ones that show no appreciable evolution from one species to another.

If we disregard anatomical variations that are directly connected to the peculiar characteristics and ways of life of particular animals, we find only one significant part of the brain that exhibits a steady enlargement from species to species: the cerebral hemispheres. In the frog, the brain hemisphere forms a swelling hardly larger than the other outgrowths from the brain stem. As we move up what we take to be the scale of animal intelligence, the hemispheres grow continuously larger. And as these cerebral hemispheres grow from one species to the next, they bulge out around the brain stem until in the ape and in man, with their cerebral cortex, they enclose and conceal most of the rest of the brain. Thus it ought to be the functions carried out by the cerebral cortex that determine man's intellectual superiority over the other animals. Conversely, we can assume that the brain stem, relatively unchanged from one species to the next at least in anatomical terms, contains centers controlling basic life functions that we share with the other animals, including the feelings of hunger and satiation, anger and fear.

Recent neurophysiological research supports the conclusions of comparative anatomy. The centers controlling not only the basic functions necessary for immediate survival but also the so-called instincts and the manifestations of emotion are located in that part of the brain we have in common with the lower animals: the brain stem.

As the cerebral hemispheres evolved from the simple cerebrum of the frog to the large cerebral cortex of man, there was a simultaneous evolution of the communications with the brain stem. Bundles of connecting nerves developed from the hypothalamus, thalamus, reticular formation, and other brain stem centers to neighboring parts of the surrounding cerebral cortex. It is as if every part of the brain stem was finally assigned its own portion of the cortex to help carry out its functions, whatever those might be. What this in fact means is that the cerebral cortex did not signify the appearance of some new qualitative capacity but only permitted the individual to perform his normal functions in a more differentiated, precise, and effective way. The fact that certain functions were transferred from lower centers to the cerebral cortex during the course of evolution does not constitute a serious objection to this view.

The General Function of the Cerebral Cortex

In looking for areas of the cerebral cortex that might have something to do with the function of intelligence, we can eliminate those that are points of arrival or departure for nerve impulses from and to the peripheries. We know that what these areas do is to receive sensory information or deliver motor instructions. Of the more than 1,000 square centimeters of surface area in the human cerebral cortex, only about one-fourth is used for such purposes. This includes the visual cortex at the very back of the brain, the sensory and motor cortices running down the sides of the two halves of the brain, plus a small area at the upper edge of the temporal lobe that serves as the terminus for auditory impressions.

Almost half of the remaining three-fourths of the cerebral surface belongs to the frontal lobes. In the course of the brain's evolution from lower to higher animals, the frontal lobes exhibit a more dramatic increase in size than any other part of the brain. The closest we can come to describing their full-time occupation is that they somehow mediate between our emotions and our intellect. It may be this process that creates motivations, and inhibitions. On a part-time basis, the frontal lobes are requisitioned by the rest of the brain to help solve difficult intellectual problems. During the last few years, a Swedish research group, using isotope techniques, has been able to designate "islands" of mental activity in the frontal lobes during various psychological tests. Thus damage to the frontal lobes can have the most serious consequences for people engaged in difficult or abstract thought—for children, who are constantly learning new concepts, and for the relatively limited number of adults who continue to learn, to think constructively, and to create.

A part of the memory seems to lie in the temporal lobes. We can produce memory reactions in patients by stimulating their temporal lobes electrically, but not by electrical stimulation of any other part of the cerebral cortex. The extensive communications system located in the *corpus callosum* makes it possible for identical memories to be stored simultaneously in both hemispheres. But memory can also be stored in the brain stem. In general, it seems to be the case that memory involving complex distinctions is stored in the cortex, while the memory needed for simpler tasks is to be found in structures located deeper inside the brain. The memory pattern for a concept or an event is probably widely distributed. The visual portion of a memory complex is in one place, the auditory in another, and so forth. Moreover, the mutual connections between related patterns in the different senses must be numerous. If an injury breaks off one of these lines of communication, there are others, at a distance from the injury, that remain, since the various senses follow separate paths between their memory banks.

On top of the fact that memory processes exhibit a wide distribution in the brain, these mechanisms also have reserves, in certain cases complete duplication. The combination of these factors—distribution and duplication—provides a satisfactory explanation for the difficulties brain specialists used to have in locating the higher psychical processes.

Several hundred years ago it was generally believed that the mind and the spirit were two completely separate things. In modern times, this dualistic view of mental and spiritual life has gradually given way to a more mechanistic point of view. We have produced machines that can carry out processes resembling thought. Every time the control over some physical process is traced to some specific part of the brain, the number of functions ascribed to the spirit is reduced, and there is a corresponding increase of what can be attributed to a machinelike brain.

It is common nowadays to stress the similarities between the human brain and electronic computers. We can now venture to say that the similarities they exhibit in their manner of operation are not merely superficial. Both computers and brains probably function the same way in the sense that they reach their similar results by essentially similar means—by a very great number of very simple operations. A computer with many working elements can accomplish more than a computer with fewer such elements. We have every reason to believe that the same is true of the brain.

Signs of Intelligence in Large-Brained Animals Other Than Man

On the basis of this discussion we can make the following assertion: animals with large brains, particularly animals with a very large cerebral cortex, can be expected to possess intelligence. One such animal is the elephant, with a brain weighing 13 pounds. Others include all the whales—the sperm whale, for example, with a brain weighing 20 pounds, and the type of dolphin exhibited in aquariums, the so-called bottlenose dolphin, with a brain weighing 3¾ pounds. Man's brain weighs 3 pounds. Before we begin looking for signs of intelligence among these large-brained animals, it may be well to try to explain what sort of intelligence we will be looking for.

When we measure the intelligence of a human being, we measure a series of abilities. Different intelligence tests measure different abilities and are therefore never comparable. Obviously, human intelligence tests cannot be applied directly to other animals, which, because of the different circumstances under which they live, have use of quite different abilities than we.

One of the qualities we regard as intelligent in such animals is their ability to generalize. This means that their manner of responding to the stimuli of their environment is not fixed or rigid. Behavior provoked by certain special circumstances can also be triggered by other similar but nevertheless different circumstances. If a circus animal has been trained to throw a ring to its trainer, then it can, without further training and without hesitation, throw a different object to its trainer. A dolphin that has learned to throw a basketball at a basket can easily be induced to throw the ball to the trainer instead, even if the basket is very close by. In other words, the trainer can become the immediate stimulus governing the action, instead of the basket. A less intelligent animal cannot generalize in this way. A chicken that has been trained to pull at a brown ring will hesitate a long time before trying its luck on a white one.

Another similarity between intelligent animals and human beings is the ability of such animals to use a pattern of behavior, a set of movements for example, within a sphere of activity completely different than the one to which it naturally belongs. In such an animal, a movement pattern normally triggered in a situation of fear or flight can instead be used in acquiring food, and vice versa. Behavior that normally has to do with sexual activity can also occur in connection with feeding. In this regard, the dolphins are masters. Virtually any combination of muscular movements can be taught and learned. Any behavior that can be triggered in the dolphin, or that the dolphin spontaneously exhibits for the trainer, can be brought under control. It is quite fantastic that dolphins, who as far as we know do not normally hunt above the surface, can be made to jump several meters out of the water and take a cigarette out of a trainer's mouth. They can be taught to juggle objects between tail and head, to vocalize in the air in imitation of human beings, and so on—apparently without end. Characteristically sexual behavior can be triggered with food in both males and females. This kind of variation in behavior, and this sort of demonstrative control over instinctive activity, are not to be found among less intelligent animals.

Intelligence and Sex

The relationship between intelligence and sexual behavior is so interesting that we will dwell on it a little. Sexologists seem to agree that as an animal stands higher on the evolutionary scale of intelligence, the part of its sexual life controlled by the lower brain centers will be smaller, and the part controlled by the cerebral cortex greater. This means that sexual behavior is freed from patterns determined in advance and transformed into a varied activity under the relative control of the individual's will, environment, and previous training.

Dolphins are remarkable in this respect. Their erections appear to be under the direct influence of the will. Six-week-old male calves can perform regular intercourse. Considering the fact that these animals reach sexual maturity at the age of five years at the earliest, this is quite extraordinary. In general, dolphins exhibit an extremely varied pattern of sexual behavior unconnected with biological productivity. Homosexuality and masturbation are practiced often by all the members of a captive colony. No individual is indifferent to the sexual advances of any one of the others, regardless of age or sex. The stimulus that triggers temporary sexual interest can be another dolphin, a different kind of toothed whale, a turtle, or any other object—including the spy-holes in the aquarium. Dolphins seem to have developed to a point where they can choose anything at all as a sexual object. They have perhaps come even further than human beings in this regard, in that we have yet to observe a dolphin becoming fixated on a biologically inadequate sexual object to the exclusion of biologically adequate objects.

A complicating factor in the study of the sexual behavior of dolphins is the difficulty of distinguishing between sex and aggression. This distinction is easy in the lower vertebrates but grows more and more difficult in the higher mammals. The one province glides over into the other, both as a concept and in the actual behavior of individual animals. Dolphins, pilot whales, and human beings often engage in what can best be described as a serious struggle between an adult male and an adult female, and then suddenly the struggle turns into copulation.

Intelligence and Neoteny

Another interesting factor in the comparison of large-brained animals is what we call neoteny, a phenomenon that runs through the development of the vertebrates like a leitmotif. The word "neoteny" designates a kind of arrested development, an evolutionary phenomenon in which certain immature, even embryonic, characteristics remain in the otherwise mature individual. The animal thought to have been the origin of the entire vertebrate series came into being in this way. This was the fishlike larva of an immobile tunicate that, through a series of mutations, became sexually mature in the larval stage and could reproduce. The fact that this new individual was free-swimming had a positive selective value, and so the mobile vertebrates gradually evolved. Another good example of neoteny is to be found in a comparison between human beings and our nearest relatives among the apes. An adult human has many features in common with a newborn ape or an ape fetus: a hairless body and a round head, among others. "Childish" characteristics such as the ability to learn and the capacity for play are more pronounced in the adult human being than in the adult of any other species of ape.

Neoteny is a process that produces an organism whose fixed muscular responses and instincts have never been locked into the adult stage. Such an organism has a great range of possible combinations and recombinations, both in determining the sequence of muscle contractions and in coupling these contractions with incoming stimuli. Neoteny produces an individual who can do all sorts of things in a great many different ways, and whose behavior in any given situation is neither rigid nor stereotyped. The ease with which dolphins and human beings can be conditioned, and the great variety of their behavior, can be seen as examples of neoteny.

Dolphins have many neotenic characteristics. An obvious example is their undisguised joy and interest in play. In fact, it was the first captive dolphins themselves who took the initiative in a great many of the simple games now played between humans and dolphins in aquariums all over the world. Dolphins, even adults, can play for hours with a ball or a feather. They are as tireless as human children in this respect.

Dolphins appear to play even in the wild state. They have been observed innumerable times riding the wakes of ships in a way that can hardly be classified as other than playful. And on two separate occasions scientists on the beach have watched dolphins surfing. On one of these occasions it could be seen quite clearly that individual dolphins returned repeatedly to the same starting point in order to catch a new wave.

Intelligence and Instinct

In connection with the concept of neoteny, we might also discuss the relationship between intelligence and instinct. If by intelligence we mean a series of capabilities appropriate to a certain animal species in certain situations—and to some extent this is the definition we use in intelligence tests—then it can be argued that the intelligence we have measured may coincide with the animal's instincts. Take as an example our attempts, in test situations, to account for the complex communications among bees about honey, or the dam-building of beavers. By means of strict instinctual control, these species find room for an astoundingly complicated program of action in a very limited cerebral space. Insects are almost completely dependent on such a solution. Their special, complex, and appropriate behavior does not admit of many alternatives. This is not what we mean by intelligence.

At the other end of the scale are the neotenic animals, not guided by instinct, who must learn a whole series of complicated procedures from their parents, and who must be prepared to solve problems on their own. These kinds of animals, whom we designate intelligent, require a long time for maturation and learning. It is often said that human beings have the longest maturation period of any animal as regards the mental faculties. It seems, however, that the elephant takes at least as long. In captivity, an elephant is not considered fit for work until it is about twenty years old. How long it takes for a sperm whale to reach the same relative maturity is not known.

Between the bees on the one hand, and human beings, elephants, and toothed whales on the other, lie all the intermediate stages imaginable. As we shall see from a comparison between the human brain and the brain of the toothed whales, it is possible that the latter is beyond or above the human brain in this matter of insight over instinct.

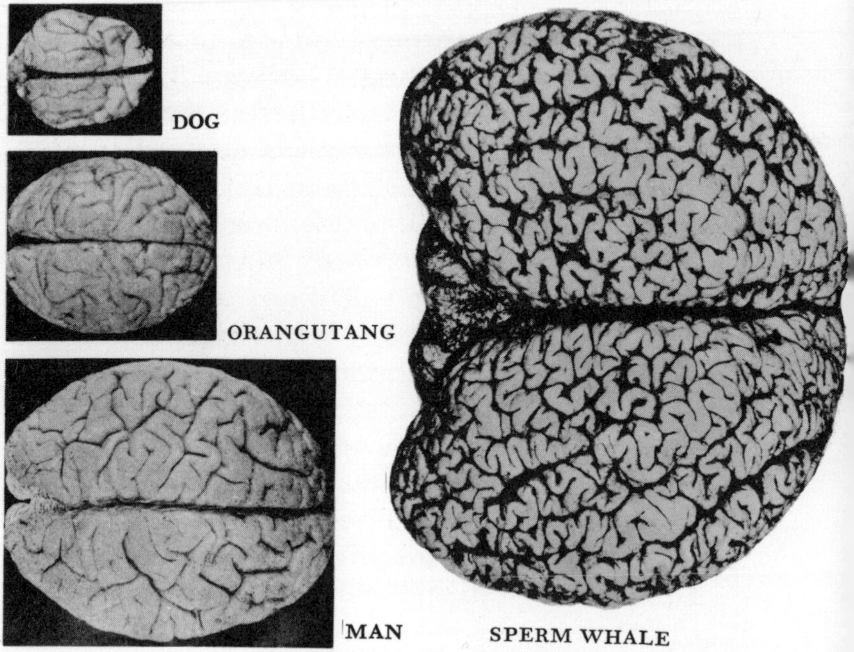

The comparative size of the brain in the dog, orangutang, human, and sperm whale. Note especially the development of the cerebrum.

Trained bottlenose dolphins in an exhibition leap.

Intelligence—an Attempted Definition

An attempt to define intelligence in accordance with the preceding discussion would look something like this: the ability to differentiate, to combine and generalize, to analyze and associate, to perceive continuity and arrive at the concept of cause and effect, to imagine the results of contemplated actions, to deliberate and find the means of reaching a desired goal.

Intelligent Behavior in Dolphins

Many stories are told about the bottlenose dolphin *(Tursiops truncatus)* which indicate this sort of intelligence. We can cite the following examples.

Two dolphins in an aquarium were playing with an eel (or at least they acted if they were playing with the eel). The eel saved itself by swimming down into a hole at the bottom. One of the dolphins then swam up to a little fish with a poison sting, took the fish carefully in its mouth, and pushed it into the eel's hole. The eel immediately fled, was caught by the dolphins, and the game went on. This episode was witnessed by the well-known Danish ethologist Holger Poulsen. A number of similar incidents have been observed.

At an aquarium in Hawaii, a team led by Dr. Gregory Bateson was able to teach a dolphin to perform new tricks, one after the other, in order to receive its reward. The dolphin not only understood what was required of it, but also showed great imagination and invention in adding to its repertoire.

Dolphins can learn by observation. In the shows given at aquariums it often happens that the dolphin expected to perform a certain trick fails to do so for some reason. It has been reported on several occasions that dolphins completely untrained for this trick have stepped in and performed it perfectly. This kind of learning by observation has otherwise been reported only in human beings and apes.

The question has been raised whether dolphins communicate with each other with the help of something that might be compared to human speech. The following experiment, which may not directly indicate intelligence, is mentioned here mostly because it has significance for the subsequent discussion. It was carried out by Dr. Jarvis Bastian in the United States.

A male and female were presented with two keys and a signal light. When the light was steady they were to depress they key on the right; when it was blinking, the key on the left. If they did it correctly—and they quickly learned to do so—they were rewarded with a fish. Then they had to learn that the male was to press first or there would be no reward. Then a screen was placed between the animals so that only the female saw the signal light. When the first signal was given, the female swam up to her keys and uttered some dolphin sounds. The male then pressed the correct key of his pair, whereupon the female pressed the correct key on her side. The dolphins could repeat this trick as often as they "wanted" to. Dr. Bastian, who performed the experiment, is himself very cautious in interpreting its results.

Intelligent Behavior in Elephants

There are also a great many stories that seem to indicate intelligence in elephants. The following two examples are taken from *Elephant Bill,* by J. H. Williams, an experienced and levelheaded observer.

An elephant with its driver was lifting large logs very high. The elephant had to balance the logs at right angles on the upper side of its trunk. Several of the logs had fallen and the work was obviously dangerous for the driver, who would have been struck by any log that fell from the maximum height. In this situation the elephant dropped a log without instructions, picked up a heavy beam lying nearby, and placed it vertically between its trunk and one tusk. When the elephant then lifted the log, the beam stood there as a protective barrier for the driver.

The same observer, J. H. Williams, gives an account, in all seriousness, of another almost unbelievable incident. A number of young elephants in captivity developed the nasty habit of stuffing mud into the wooden bells they wore around their necks so that they could no longer ring. Then they stole bananas during the night. In this way they managed to plunder entire banana groves in the immediate vicinity of the plantation owner's house.

In 1957, Bernard Rensch described the following experiment performed on a five-year-old elephant in a zoo. It was taught to pick out one of two visual symbols in order to get food. It learned to remember 20 such pairs of symbols at one time, and presumably could have learned more. As the process continued, it learned faster and faster. One year later, the same elephant was tested on 13 of these pairs and managed all of them but one, which had been difficult to learn in the first place.

Can Animals Without a Culture Like Ours Be Intelligent?

We cannot yet say how intelligent dolphins and elephants are, because we do not know what they use their large brains for. The only thing we can say for certain about these large brains is that in some way these animals must have found them, and still do find them, very useful. Function, or behavior, is a leading factor in evolution. This implies that if dolphins and elephant use their big brains for thinking, then thinking has had survival value for them.

One very superficial argument against the suggestion that the other large-brained animals might measure up to us in regards to the higher mental processes is that these animals have

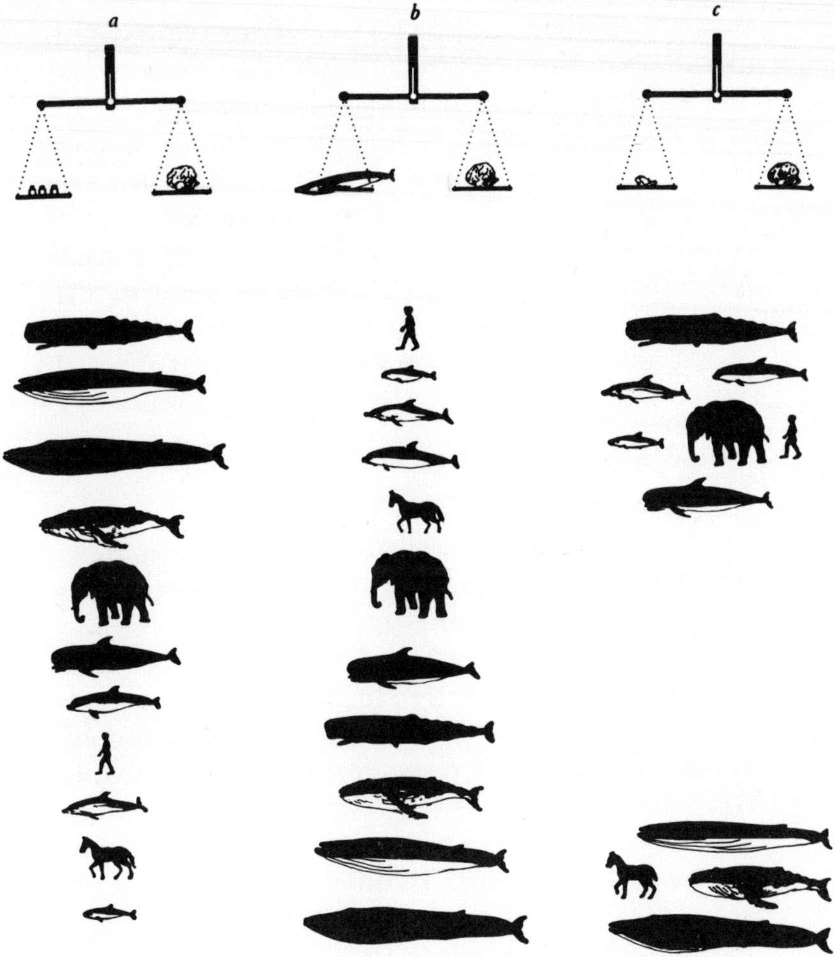

Ranking of a number of animals according to (a) total brain weight, (b) relation of brain weight to body weight, and (c) relation of brain weight to brain stem weight. The sperm whale tops columns (a) and (c). The diagram incorrectly shows man topping column (b)—it should be the marmoset or possibly the tiny shrew.

never managed to develop a culture equivalent to our own. We must not forget that *Homo sapiens* had had his large brain and its potential for at least 35,000 years before he suddenly began, maybe 10,000 years ago, to cultivate the soil and develop a written language, with the subsequent rapidly accelerating technology. It is hardly likely that this leap forward in cultural evolution was accompanied by equally dramatic changes in mankind's mental potential.

The Human and the Dolphin Brain Compared

The literature that exists on the relationship between the brain's appearance and intelligence is not particularly sophisticated. Human beings have simply assumed that they were the most intelligent and have then looked for the morphological support for that assumption. But the assumption itself may be wrong. It is very probable that we are the most intelligent, if by intelligence we mean those abilities we can measure with our present intelligence tests. But it is far from certain that we are the most intelligent, if by intelligence we mean, say, the capacity to alter behavior into a number of different patterns in response to changing information from the environment.

At first we explained our supposedly superior intelligence by the fact that we had a larger brain than other animals. When it then became clear that there were large animals that had larger brains than we did—elephants and whales—it was easy to carry out our first retreat and say that of course it was brain weight per kilogram of bodyweight that ought to be counted. But in that case, *Homo sapiens* comes in second to the marmoset. All sorts of strange indices involving brain weight were worked out, but have been discarded. As for the direct comparison of brains, we discovered under the microscope, to our joy, that the nerve cells were farther apart in those animals with larger and more convoluted brains than ours. But our joy was not quite so great when we discovered it was a general rule that brain cells become less compact as the brain becomes larger and more complex. The most recent thing we have latched onto is that the six layers in the human brain are architectonically better arranged than in the dolphin brain. To maintain that doubtful differences in the brain's capacity lie concealed in equally doubtful dissimilarities of a morphological kind cannot be called good science.

There seems to exist a certain correlation between brain size and body size in closely related animals. It is reasonable, after all, that it should take a greater number of nerve cells, and most of all longer nerve cells, to keep track of a larger body than of a smaller one. But a large body cannot need so awfully many more, particularly if a great part of it consists of fatty tissue, as in the whale. The truth of the matter is that we must think of the brain as a biocomputer and then admit that it is probably the total size of the cerebrum that correlates with intelligence.

The human brain weighs 1,400 grams, and the dolphin's 1,700. The dolphin's cerebral cortex is larger than ours. It has twice the number of convolutions, and 10 to 40 percent more nerve cells. The mapping of the dolphin's cerebral cortex is going on at the present time. The dolphin has a larger acoustical cortex than we, roughly equivalent to our visual cortex. But the human motor cortex is considerably larger than the dolphin's. On balance, then, it looks as though the dolphin has more cortex left over for the higher mental processes than we do.

Vision and Hearing in Dolphin and Man

Everything that reaches or leaves our brains moves through multiple channels. For each of our eyes there are 1,200,000 neurons running between the retina and the visual cortex. All of these operate in parallel, and each channel delivers information at a rate of about 50 bits of physical information per second. In spite of the relatively slow input of each channel separately, the total rate of information received by means of vision is considerable. These figures are cited by Dr. John Lilly in his book *The Mind of the Dolphin.* Dr. Lilly notes in the same book that the dolphin's visual input amounts to only one-tenth of our own.

Every human ear is connected to the brain by 50,000 nerve fibers. So we are a long way from

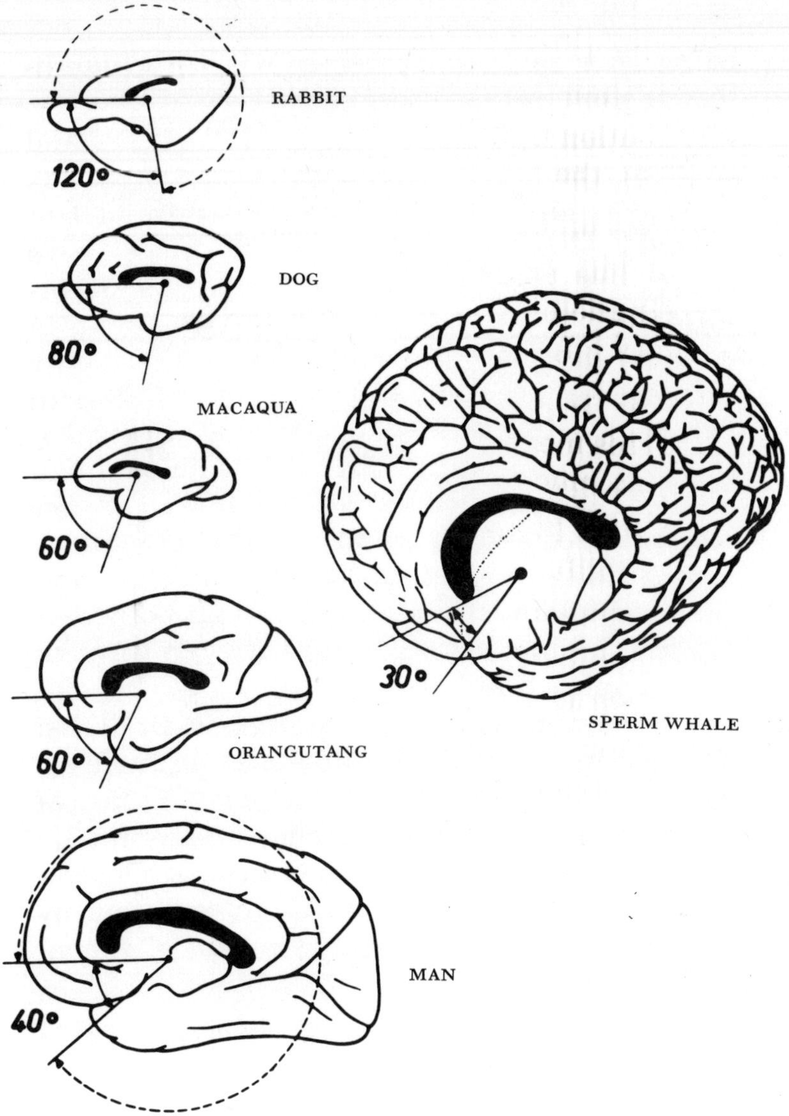

One way of describing the degree of cerebral development: the less the relative distance from the lower portion of the frontal lobe to the upper limit of the temporal lobe, the more highly developed the brain. As we can see, this distance is smallest in the sperm whale.

being able to receive as much information through our ears as through our eyes. The dolphin has 2 1/4 times as many nerve fibers from each ear as we have—115,000 fibers. Furthermore, the dolphin's hearing functions at frequencies that are on the average 4 1/2 times higher than those we use. According to physical information theory, the higher the frequency of a signal, the more physical information it contains per unit of time, and the greater the number of bits transmitted per second. Logically, then, the dolphin can receive 4 1/2 times as much auditory information per second as we can, because its frequencies are 4 1/2 times higher than ours. This means that the dolphin can take in 20 times as much information through both ears (40 million bits per second) as we can through ours (2 million bits per second).

In summary, then, according to Dr. Lilly: The dolphin receives only one-tenth as much information as we do by means of vision. By means of hearing, on the other hand, we take in only one-twentieth of what the dolphin does. The dolphin obtains almost as much information by means of hearing as we do by means of vision.

The Limbic System in Large-Brained Animals

There is an interesting quantitative difference between the brains of the toothed whales, on the one hand, and the land-dwelling mammals, including man, on the other. The neocortex, in evolutionary terms the youngest part of the cerebral cortex, is comparatively better developed in certain toothed whales than in man.

Another interesting difference between the brains of the toothed whales and the brains of human beings and elephants involves the so-called limbic system. This is located in the middle of the brain and consists partly of the cortex on the inside of the two cerebral hemispheres. One part of the limbic system is thought to store memory, another part is supposed to decide which memories are to be stored, and a third part takes care of the communications between these and other parts of the brain. The system as a whole is presumed to choose the appropriate behavior for different groups of incoming stimuli. In this regard, the remainder of the cerebral cortex can be looked upon as a computer system connected to the limbic system. The cerebral cortex as a whole takes care of problems too difficult for lower parts of the brain to cope with, while at the same time it maintains a permanent memory bank and sees to certain special higher functions such as speech.

This view of the limbic system implies that a very important part of our emotional life must originate there. The limbic system may contain the reward pathway that we have good reason to believe is built into the central nervous system as an aid to learning. And from this point of view, it is extremely interesting to note that the toothed whales have a larger remaining cerebral cortex in relation to their limbic system than we do. Does this imply that the toothed whales are governed less by emotion and more by reason than we are?

Can the Dolphin Brain Be Superior to Ours?

The surprising conclusion of this comparision is that the dolphin brain *could* be superior to ours. It is also possible that we have overlooked something important. Because it *could* be that the dolphin brain is much poorer functionally than ours. There might be biochemical differences we know nothing about, which could mean that the speed of the neural processes in dolphins is of a very different order. Most neurophysiologists consider such differences unlikely, however.

One more such "pessimistic" note. As will be noted further on, it is probable that dolphins have a well-developed acoustical memory. It could be that acoustical memory requires a greater number of neurons than, say, visual memory. This would mean that a large portion of the dolphin's cerebral cortex consisted of acoustical memory cortex, and that a relatively smaller part would be left over for the other higher mental processes. But about this we know nothing. The dolphin brain continues to be a great challenge.

A Medical View of the Human Cerebral Cortex

It might be constructive to conclude our comparison of the human and dolphin brains with some reflections on the human brain and its workings by a professor of neurology who seems to be unaware of the existence of the other large-brained animals. The following is from a book on the function of the brain, written by Mogens Fog in 1964.

> Considering the development and capacity of the human brain, it is not without a certain justice that man is regarded as the master of creation. Only in this respect do we differ markedly from the animals.
>
> The higher an animal stands on the evolutionary scale, as man reckons it, the greater the area occupied by its cerebral cortex in relation to the central cell masses, the basal ganglia. The surface area of the cortex not only increases as the reslut of more space in larger craniums but, to an equal degree, as the result of a greater number of convolutions separated by fissures that become steadily deeper. In order to explain this change in volume we must compare the skills of various animals. A large cortical surface permits a greater power of combination and better opportunities of acquiring experience, which is gathered through the various sense apparatus and tied together into a comprehensive picture. Thus it allows a subtler and more purposeful relationship to

> the surrounding world. Let us call it greater intelligence.
>
> The function of the cerebral cortex is to make fine distinctions, preserve them in a memory pattern, and let a greater and greater degree of evaluation determine procedure. A cat obviously makes certain deliberations before it decides whether to jump from a tree limb to the ground. The larger the cortical area, or in other words, the greater the number of available neurons, the more sophisticated the background to a decision, a shall—shall not.

I believe Fog's view of the function of the human brain is by and large correct. But what then is the function of the dolphin brain? Until proven otherwise, it is scientifically sound to apply the same principles in assessing the dolphin brain as we do in assessing the human brain.

A Cetologist's View of the Brain of the Whale

Mogens Fog is a specialist on the human brain and has not given much thought, presumably, to the brains of other animals. But what does a specialist on the whale brain have to say? After all, he must make comparisons with the human brain pretty much the way I have done. The Norwegian Jan Jansen is such a specialist. In an article published in 1969 he gives an excellent discussion of the characteristics of the whale brain, and states that continued research is very desirable. But he would give priority to studies of the neural patterns connected with motor and acoustical activity. He maintains that a discussion of the intellectual capacities of the whale is premature, and contents himself with noting that observations made on captive toothed whales indicate a considerable capacity to learn. He places the bottlenose dolphin somewhere between the chimpanzee and the dog.

In our view, however, the simplest anatomical comparison between the human brain and the brain of the toothed whale tells us that the most important research remaining to be done on the nervous system of the toothed whale concerns the function of the cerebral cortex. Such studies are no more premature than similar studies on human beings.

The Evolution of the Human Brain

We like to believe that we developed our large brain so that we could read and write and reckon, but as we have already pointed out, prehistoric man, who could do none of these things, had a brain just as large as the one we have today. Of course it could be that the brain of prehistoric man was inferior to ours in some functionally essential way, even if it was just as large. In all probability, however, their brains had almost exactly the same potential as ours. Which is not to say that these prehistoric people were just as "smart" as we are. For we have been able to demonstrate that people can *learn* to think. Since this has great significance for an appreciation of the potential of other large-brained animals, as well as for an understanding of the evolution of our own brain, we will describe these experiments.

Thinking Can Be Learned

Innumerable "learning" experiments of the trial-and-error type have been conducted with animals. Harry and Margaret Harlow wondered if, by means of such experiments, apes could be induced to solve problems immediately, through insight. The apes were exposed to a series of so-called discrimination tests. Two objects of unlike color, size, and shape were placed on a small table. The ape was rewarded if it picked up the right object. The experiment was repeated over and over again with several hundred pairs of different objects. In this way the ape was presented with numerous problems, all of the same general type. It was only the objects that changed.

In the first attempts, the ape solved the problem by means of the tedious trial-and-error method, but as it solved problem after problem of the same kind its behavior began to change dramatically. It solved every new problem with greater and greater ease, until at last it showed perfect insight. If it chose the "right" object on the first attempt it made no further mistakes, and if it chose "wrong" the first time it chose "right" the next, and the next.

The Harlows believe this experiment shows

that trial-and-error and insight are two different phases of a long, continuing learning process. They are not two separate capacities, but steps in the process of learning and thinking.

Children between the ages of two and five reacted just like the apes. As a group, the children learned more quickly than the apes, but they made the same kinds of mistakes. The "smartest" apes learned faster than the "dumbest" children. Experiments that were more complicated than the one cited were carried out with essentially the same results. All the studies showed that the capacity to solve problems correctly is not inborn but gradually acquired. This conclusion appears to apply to human beings too.

The Harlows call this process of progressive learning the formation of a "learning set." An individual learns an organized set of practices that helps him deal with every new problem of a particular kind. A single learning set gives an animal only limited help in adapting itself to a constantly changing environment. But a very large number of learning sets can make up the raw material for human thought.

Over the years, a human being learns to deal with more and more difficult problems. At the highest stage of this progression, an intelligent, full-grown human being selects the raw material for his thinking from innumerable previously acquired learning sets. His many years of education in and outside of school have been devoted to the formation of these complex learning sets, which he eventually uses with such ease that he and those who study him can lose sight of their origin and development. It is easy to understand the significance of spoken language for the production of human learning sets, and to appreciate what an enormous step forward the invention of writing was, in the sense that learning sets acquired through bitter experience could be passed effectively from one generation to another.

Principles of Biological Evolution

After this introduction to learning sets, we can approach the question of the evolution of man's large brain. But first we must discuss, in a general way, the principles that apply to biological evolution. Above all, it is important to understand the way in which evolution is affected by behavior, or function. We can illustrate this with an example from Robert Ardrey's *The Territorial Imperative.*

For millions of years, a species of small rodent has lived on a grassy plain. The grass has given it both protection and food. A change in climate then occurs. Decade after decade, century after century, bring drought. The grass dries up and the landscape, once so lush, turns into a desert. The little rodent must alter its entire way of life or die out. It must eat new kinds of food, and hoard and invent new ways of preserving the food it previously had in overabundance the year round. By natural selection over the course of hundreds of generations, those animals with a tendency to hoard have greater success in feeding their young than those who tend to live for the moment. Hoarding behavior is established, and that leads immediately to a need for storage space. The little rodent has to dig. And this new burrowing life means that the pressure of natural selection favors those individuals whom chance has equipped with paws that are a little better at digging than those of the competition. The light-footed rodent becomes a new species with heavy feet and claws.

The life of a desert burrower presented challenges of a different kind as well. From the beginning of its new career, the rodent was exposed to the hawk. There was no longer any grass to hide in. The dark rodent from the grassy plain became the hawk's favorite dish. The new desert rodent developed protective coloration and became hard to see from the air. But hawks are capable animals and many rodents were spotted from the air despite their protective coloring. The burrow became a hiding place. It also became a shelter from the burning summer sun that attacked both the animals and their food supply. During the merciless summers, the little rodent spent more and more of its time in its burrow, asleep. Physiological changes followed the changes in behavior, and the rodent became a summer hibernator the way the bear became a winter hibernator.

And so on. Birds do not fly because they have wings. They have wings because they fly.

We Have Large Brains Because We Think

In all probability, the same basic principles that applied to the evolution of the rodent's front paws also applied to the evolution of the human brain. Being able to think has been useful to us, and every improvement in the functioning of the brain has been rewarded in the struggle for survival. In other words, we have a large brain because we think. Not, we think because we have a large brain. That seems obvious to a biologist. But it does not seem to be obvious to philosophers. Arthur Koestler, who is very interested in biology, maintained as recently as 1967 that the swift evolution of the brain meant that our ancestors were equipped with a luxury organ whose capacity far surpassed its owner's immediate primative needs. Since the belief in this overcapacity of the brain seems to be so deeply rooted, we might examine it more closely.

Koestler uses an analogy to make his point clear. An illiterate shop owner in an Arabian village, we can call him Ali, was poor in arithmetic and was often cheated by his customers instead of the other way around. Every evening he prayed to Allah for an abacus. But something went wrong with Allah's communications, and one morning Ali found his shop filled with the latest IBM computer, its instrument panels covering the walls. After twisting different knobs for several days without result, Ali lost his patience and started kicking the machine. This disturbed one of its millions of electrical circuits, and after a time Ali discovered to his joy that if he kicked a certain panel three times and then five times, one of the dials showed eight. He thanked Allah for giving him such a wonderful abacus, and went on using the machine for simple addition—blissfully unaware of the fact that it was capable of deriving Einstein's equations in a few seconds, or of predicting the movements of the stars and planets for thousands of years to come.

Ali's children and then his grandchildren inherited the machine and the secret of kicking the panel. It took hundreds of generations before Ali's offspring learned to use it for even as simple a thing as ordinary multiplication.

We ourselves are Ali's descendants, and even if we have discovered many other ways of getting the computer to work, we have still, according to Koestler, only learned to use a small fraction of the potential of the hundred thousand million circuits the human brain is thought to represent.

Koestler understands that evolution provides for the immediate adaptive needs of the species. He realizes that it would be biologically unique for evolution to equip a species with an organ the species did not know how to use, a luxury organ like Ali's computer, which many times surpassed its owner's immediate primitive requirements. And still he comes to the conclusion that this is the case with man and his brain.

A scientist trained in biology must assume that every alteration in the brain's construction, and thus in its function, must have had immediate survival value for the species. It is highly probable that modern man makes better use of his brain than Cro-Magnon man did of his, though it had the same size and form. But in order to explain this we needn't have recourse to the theory of mental overcapacity, a theory that is contrary to the doctrine of evolution. The increased number of learning sets to which a modern human being is exposed from early childhood onward is quite an adequate explanation. Our most recent great achievement in the presentation of learning sets, television, has already had great significance and will have even more in the future. Anyone now middle-aged or older will have to admit that young people have an advantage in that, if they survive, they will probably be, on the average, smarter and better equipped to meet the problems of the modern age.

Those students of the mass media who have a hard time demonstrating the effect of the media in some particular context might profit from looking at their research projects in this larger perspective. It is obvious that in the long view the mass media are enormously influential. It may be difficult to demonstrate this influence in short-term experiments, but that is another matter. The way we program our built-in computers can never be considered unimportant.

The learning sets of Cro-Magnon man were

suited to Stone Age problems. Is modern man's problem-solving ability equally well adapted to the problems he meets? Can we expand our learning sets to deal with the problems of progress as quickly as we create them? Is the total capacity of the brain great enough to deal with the increasing number of learning sets we will have to acquire in order to prevent the catastrophes that threaten us? We will come back to these questions in the final chapter.

The Mechanisms of Selection Behind the Evolution of the Human Brain

So far, the discussion leads us to the conclusion that the human brain evolved in essentially the same way as all other organs, that is, gradually, led by function and behavior, and that every small advance had at least temporary survival value. All of these improvements have helped us slowly—and in some cases suddenly, by means of new inventions—to increase the number of learning sets to which an individual is exposed during his lifetime. This in turn has led to the brain being used in a way that was not originally "intended." The development of civilization has increased or decreased the selection value of old improvements.

But it is not at all clear which selective pressures and which improvements have been most significant for cerebral evolution in general. All we can do here is guess, estimate, and suppose.

We can base our speculation on two different kinds of data—on the fossil remains of our ancestors and their relatives, and on studies of analogous characteristics found in living primates and other mammals. It is impossible to interpret fossil finds without making reference to the behavior of living mammals, including man. M. R. A. Chance and A. P. Mead have helped to define the problem by identifying two aspects of primate behavior that they consider unique. Among primates in whom the female is sexually receptive for more than one-third of her menstrual cycle, copulation is possible over long periods of time and thus is just as important as other forms of behavior within the group. This situation gives rise to an element of conflict that affects all other group activity. This in turn results in a unique type of natural selection.

Carpenter and others have demonstrated that not all males in the different subhuman primate societies are equally active sexually. In many such societies, males who fail to achieve sufficient social status never get a chance to breed. It's a question of belonging to the ingroup, where the breeding prize is dealt out by the leading clique. And in order to please the leaders, the young male must be clever, and must be able to control his aggression in certain situations—which certainly requires some of those higher mental capacities we ascribe to the cerebral cortex of modern man.

The Development of Language and the Evolution of the Cerebral Cortex

It is possible that the individual's need to adapt to the group has had great importance for the evolution of the cerebral cortex. But since everything points more and more to the supposition that natural selection affects primarily groups of animals rather than individuals, perhaps we should ask ourselves what kind of group behavior has encouraged the higher mental faculties. In which case it is quite evident that the complicated teamwork required by hunting and war must have had considerable significance in the course of the 10 to 20 million years our ancestors are thought to have been hunters. There are no other hunting apes. We would especially like to wager that the development of language and of the cerebral cortex has gone hand in hand, and that it is the linguistic function that has meant most for the evolution of the cerebral cortex in *Homo sapiens.* Scientists conjecture that complex language was used by manlike creatures who lived almost 2 million years ago and who has a brain slightly larger than that of the chimpanzee. It must have been a tremendous advantage for the members of a tribe to be able to communicate with each other and discuss their plans before a hunt or a battle with their neighbors, and then to be able to give orders and information during the struggle itself.

The Production of Tools

Once cerebral evolution had gotten under way and something of a culture had come into being, new factors appeared that doubtless accelerated the development of a large brain. The same manlike creatures that are supposed to have had a language, *Australopithecus,* were also the first large-scale producers of tools. Ape-man's tree-climbing period was a prerequisite for tool production because it was then that he developed his prehensile hand. Cultural evolution is nonbiological in the sense that it is passed on to every member of the culture in question regardless of his or her genetic origin. But otherwise, cultural evolution is an aspect of biology. Our own culture has come to be one of the strongest selective pressures ever, in a class with the great climatic changes. Genius and creativity play an important role in biology. The group that wins the struggle will be the one whose average intelligence and originality permit the random appearance of creative individuals of a higher vitality than those in the competing group. It is fruitless to talk about natural selection and at the same time ignore the influence of culture.

"Cultural Mutations"

It is not easy to imagine or prove the occurrence of "cultural mutations" in prehistoric societies, that is, cultural improvements that might have had importance in the struggle between two tribes. Nevertheless, here is an example that is quite impressive. Certain Bedouin tribes are said to have been in the practice of putting stones in the uteruses of their female camels. Presumably, these stones had the same effect as a present-day coil and prevented the she-camel from bearing young. A pregnant camel could unquestionably be a great burden to a Bedouin tribe during long treks in times of drought. The tribe that included an individual smart enough to make this fundamental innovation—however it may have come about—was in a better position than the tribe that could not "improve" its environment in this way. The first group could extend its travels when the circumstances demanded it. Besides being physically handicapped, a pregnant she-camel is also bad-tempered. A camel bite is said to be very nasty. One or two men unfit for battle, or even dead, because of camel bites could decide against the tribe in a battle with its neighbors.

Impressive examples of cultural traditions that are both meaningful and insightful have been found in Stone Age peoples living today. Thus the Papuans produce salt from a salt-rich reed. First they burn the reed to ashes. Then they extract the salt from the ash with water. Then they recover the salt from the water by evaporation. In a tropical climate such as the Papuans live in, it is highly probable that the tribe that could produce salt won out over the tribe that could not. Among other things, lack of salt can cause cramp during exertion at high temperatures.

The most important "cultural mutation" to date is probably the invention of writing. We emphasized its importance for our culture in the first chapter.

Even in modern times there are examples of the way natural selection between large human groups depends directly on which "cultural mutations" precede the struggle. An obvious case is the extermination of the North American Indian. Open violence need not accompany the death of a culture or a people. The threat of violence is enough, when one party is sufficiently superior. In Sweden, the Lappish culture has gradually been pushed back and with it the relative proportion of the population that was Lappish. With the incorporation of Lappish culture into Swedish society, the gene pool of the Swedish people becomes slightly altered, but the unique gene pool of the Lapps disappears entirely.

The Largest Part of the Functioning Human Brain Lies Outside the Body

In discussing the evolution of the human brain it is important to remember that the final result, the brain of the individual modern man, is not as remarkable as we like to believe. As a result of cultural evolution, and particularly as a result of the invention of writing, the greatest part of its capacity lies outside the individual. The combined mental capacity of modern society is enormous. By comparison, the capacity of any given genius is quite small. In almost everything we do in our daily lives we are dependent on the experience of earlier generations and on the help of our fellow men. If something goes wrong with the car, we resort first to the experience we have acquired through other people, then we turn to handbooks and garages. And even at that, we begin with something as complicated as an entire automobile. Think if we had to discover, invent, and make everything that goes into an automobile. We wouldn't get very far in one lifetime. The largest part of a human brain operating in a modern society can be said to lie outside the body. What the individual accomplishes with the power of his own intellect is not especially remarkable; in most cases it is embarrassingly paltry in comparison to what we give ourselves credit for.

For What Purpose Do Dolphins Use Their Large Brains?

All attempts to answer the question of what dolphins use their large brains for must start with the fact that the brain evolved through natural selection like all other organs, and that function led the way. This means that we must first of all try to acquaint ourselves with what the dolphin's life in the sea is like. A dolphin in a free state uses sonar with a long wavelength for purposes of general orientation. It uses shortwave sonar for detailed analysis of interesting objects. Theoretical calculations by Dr. Seville Chapman at the Cornell Aeronautical Laboratory in Buffalo, New York, indicate that the dolphin should be able to achieve the same resolution with its sonar as we can with our vision. We can see particles as small as 0.1 to 0.2 millimeters in diameter. Dolphins should be able to hear particles equally small. According to Dr. Lilly, dolphins receive roughly as much information with their ears as we do with our eyes. When we combine sight and hearing, the result is that dolphins can take in approximately as much information as we can.

One clear advantage the dolphin has over us is that it can, in all probability, listen *through* objects. Sound waves in water penetrate a dolphin or a human being with only the loss of a little reflection and absorption. Skin, muscle, and fat are virtually transparent to sound waves carried by water. Air cavities and bone, on the other hand, reflect sound waves relatively strongly—the air cavities in particular. A dolphin "listening" to another dolphin hears the body contours diffusely, teeth and bone somewhat better, and those parts containing air—the alimentary canal, the breathing passages including the lungs, and the air cavities in the skull—quite distinctly. The dolphin hears in a way that resembles the way a roentgenologist uses X-rays to see. (To help the reader imagine what the dolphin hears, we have included a picture of

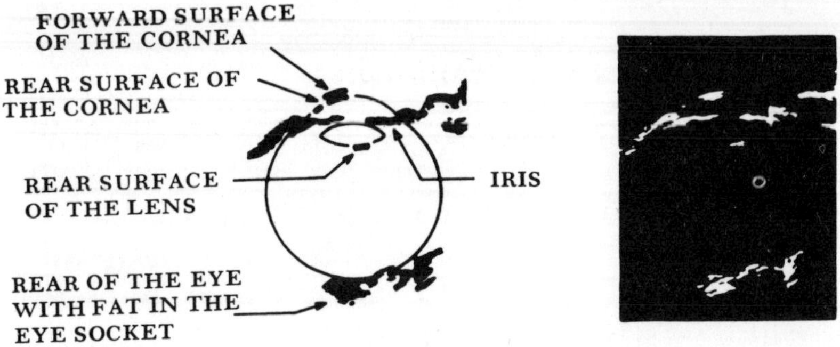

An ultrasonic picture of a living human eye. The drawing on the left shows the echoes in relation to the eye's structure.

the human eye, produced by a complicated apparatus using only sound waves.)

This must have considerable significance for the communication of emotional states and thus for all personal interaction. We can imagine one dolphin saying to another, "Darling, I can hear that something's upset you. You're swallowing air." Dolphins probably have a hard time dissembling. We express a part of our feelings with our facial muscles, which can be consciously controlled. Dolphins reveal themselves, among other ways, through intestinal peristalsis, which is not directly under the influence of the will. It is impossible for one dolphin to play tough with another. People who have difficulty communicating their feelings will easily appreciate the advantage that dolphins enjoy in this respect.

Large Acoustical Memory

The size of the dolphin's brain, combined with its exceptionally highly developed hearing apparatus, indicates that it should have a large acoustical memory. The dolphin may need an acoustical memory if it has a language, but we will return to this important question in a later section.

Dolphins range over a large area of ocean and could make good use of the ability to recognize their surroundings with their longwave sonar. We know that several species of whales meet annually in specific locations in order to breed, but we do not know how most of these whales find these breeding grounds. It could be that some of them are dependent on acoustical memory for finding their way. In which case those who have a bad memory for the sound pictures of the sea never get the chance to reproduce.

However, it is not certain that the acoustical memory of whales affects natural selection so directly. We know that eels can find the Sargasso Sea without having learned the way. On the other hand, the situation for aquatic mammals is different than it is for those animals that have always been adapted to a life in the sea. Fish and similar animals that evolved in water from the beginning maneuver by means of reflex motion patterns that demand no intelligence. When certain fish began living on land and, via reptiles, evolved into mammals, these reflex patterns were lost and thus did not exist in the ancestors of whales when they returned to the sea. They were already quite highly developed mammals with considerable intelligence, and they had to *learn* to dive and to orient themselves in water. We would like to believe that this put a high premium on intelligence and the ability to learn —a sort of parallel to the way man, at least partly through intelligence, was able to conquer that part of his environment previously dominated solely by the large predators through innate morphological and behavioral adaptations. The dolphin learned to be a "fish" the same way primitive man learned to be a "wolf." Whales could make up for the eel's innate ability to find his spawning grounds by learning to recognize and to remember sound pictures of the sea.

But as we said, such speculation is very risky. The first whale, which already showed an extensive adaptation to aquatic life, had a comparatively small and probably rather smooth cerebral cortex. The manatee, which is well adapted to the water, has a brain no larger than that of its fossil ancestors. The beaver, finally, which looks to us like a clever engineer, is governed by instinct to an astonishing degree. So it does not seem that aquatic life necessarily puts a high premium on the ability to learn.

The literature on dolphins contains a weakly supported supposition to the effect that the use of echo-sounding presupposes a certain amount of learning for each individual. Echo-sounding animals would then require a cerebrum of considerable proportions. This is not necessarily the case, however, as is demonstrated by the bat, which is a master of echosounding but equipped with a very modest brain.

Aggression and Friendship

Another piece of conjecture about the dolphin's use of its brain, quite as risky as the first, concerns the relationship between aggression and friendship.

Bonds of friendship between individual members of a species can be studied in about the same way we study sight and hearing. Such friendship bonds make their first appearance in the animal kingdom in several very aggressive bony fishes. In birds, the strongest bonds of friendship occur in the aggressive goose. Among mammals, pronounced bonds of friendship exist in certain aggressive species—in wolves and human beings, to name a couple of good examples. There is thought to be a direct connection between aggressive behavior and the ability to establish friendships. If nothing else, it has been convincingly demonstrated that aggression in the daily social life of animals—not only during the breeding season—is a necessary prerequisite to the establishment of friendships between individuals.

Dolphins are a question mark in this regard. There is no mistaking the friendships they make among themselves. There are moving examples of dolphin sorrow and dolphin joy when some good friend or mate disappears or returns. Like human beings, dolphins can develop ulcers from grief.

Dolphins are never aggressive toward human beings—not even in captivity. That they never lose their patience and use their sharp teeth to bite someone is in fact very odd. Even the so-called killer whale (the largest of all dolphins with a brain weighing almost 9 pounds), known for its healthy appetite for all types of animal protein, is, in captivity, friendly toward human beings. And yet the captive whale can hardly take a positive view of its situation. An ordinary display tank must be to its occupant what solitary confinement in a pitch-black cell would be to us. In every direction, its sound signals bounce back from cement walls, just as the eye of the human captive registers nothing but blackness. One wonders if this control of aggression can be a function of the large brain and whether it has been of importance for the survival of the species. Far too little is known about the way the dolphin expresses aggression in the wild state for us to come to any conclusions on this point.

The dolphin's spontaneous tameness and its ability to control aggression in captivity are interesting in connection with Chance and Mead's theory that young male primates adapt their social behavior to that of the dominant males, and that this adaptation has had significance for the evolution of the brain.

Natural Selection Applies Primarily to the Group, Not the Individual

In the section on the development of the human brain we maintained that natural selection applies primarily to groups of animals rather than to individuals. Evolution can supply innumerable examples of the individual's interests being set aside in favor of those of the group. We know that hunting dolphins work with each other and even with human beings in an apparently intelligent way. We can cite the following example.

In 1954, an American specialist in tropical forestry named Bruce Lamb described a case of a freshwater dolphin, the buoto, helping men fish. It was night fishing. Upon reaching the fishing spot, the man paddling began to whistle and beat his paddle against the side of the canoe. Lamb was told that they were calling the buoto. One of the fishermen lit his carbide lamp, chose a harpoon, and took his place in the bow, ready to start the fishing. Almost immediately afterwards, a dolphin surfaced to breathe about fifty feet from the canoe. The canoe was paddled toward shallow water near the shore and the first fish was speared. The school broke up and the fish headed for deeper water. But there they ran into the dolphin, who in all probability was doing some fishing himself. In this way the fish were chased back to shallow water and the harpooning could continue. Now and then the dolphin could be heard surfacing and breathing out in the darkness. The canoe drifted slowly downstream, and the dolphin maintained his position the entire time. Lamb's companions assured him that they always had the help of the same dolphin when they fished at night. If it was not in fact the same individual, they did in any case have the help of dolphins.

Several stories concerning the teamwork of dolphins with each other have been reported by American dolphin experts. On one occasion, for example, a school of dolphins was observed approaching an artificial obstacle in the open sea. A scout separated from the school made several echo-soundings of the obstacle, found a passage, returned to the others, and piloted them through.

There are also many stories of how dolphins help unconscious companions who are in danger of drowning, both in the wild state and in captivity. There is not the least doubt but what on certain occasions dolphins behave purposefully. Cases of something very much like heart massage have been described.

Absence of a Culture Resembling Our Own Does Not Rule Out Intelligence

Dolphins have no civilization comparable to ours. The prerequisite for our culture is the combination of a large brain and a prehensile hand. The dolphin has the large brain but not the hand. It never had the tree-climbing period during which the prehensile hand evolved. Its ancestors went into the sea perhaps 50 million years ago. These dolphins-to-be managed to survive the pressures of natural selection without developing a culture. We do not know exactly what the selective pressures were that affected these land animals and resulted in the evolution of a large brain. We can only guess. The fact that they lack hands obviously does not mean that, in the course of their evolution, they had no use for a brain that in many respects resembles ours, and surpasses ours in quantity. The large brain of the dolphin continues to be a challenge.

Do Dolphins Have a Language?

In earlier sections we maintained that the evolution of human language went hand in hand with the evolution of the cerebrum, and that the need for language must have governed this development. An important, enthralling question now follows naturally: Do dolphins, and whales in general, have something that might be called a language?

There is much to indicate that in mammals there is a critical absolute brain size below which language is impossible and above which language is not only possible but even probable. Modern information theory holds that the number of interconnecting active elements determines functional capacity. If the number of elements in a computer falls beneath a certain figure, there are certain things the computer cannot do. The normal speaking and writing human brain contains 13 billion interconnecting neurons and 65 billion glia cells. A modern computer contains significantly fewer (theoretically equivalent) elements. And in regard to the number of elements, the dolphin brain is superior to the human.

Another requirement for language is that there be something to communicate. As previously stated, neuroanatomical research has shown that the dolphin can take in as much information by means of its vision and hearing as we can. Like us, they are social animals and ought to have as much to communicate as prehistoric man, who very definitely had a language.

A third prerequisite is of course some means of expression. The dolphin is certainly not deficient here.

Sound Production in the Dolphin

Sound production in dolphins has been studied intensively by American scientists for the last twenty years. Numberous recordings have been made of wild dolphins, but the studies have concentrated primarily on dolphins in captivity. Yet we still have no firm picture of dolphin sound production, and I will confine myself to giving an account of the work of two of the most well-established specialists, the Caldwells, a husband-and-wife team.

According to the Caldwells, the vocalizations produced by dolphins can be divided into two groups: (1) pure tones, generally called whistlings, and (2) sounds composed of rapid throbs, the so-called pulsing sounds. The latter are further divided into the sounds used in echo-location and the sounds the Caldwells claim are used in emotional contexts.

As for the whistling, the Caldwells consider it highly improbable that it is used in any kind of complex dolphin language. They have discovered that every dolphin has its own characteristic whistle outline, its own whistle signature. The number of whistling sounds a dolphin emits per unit of time can vary considerably, as can strength and duration, but the basic character of the sound in each individual shows very little variation. The personal nature of the whistling means that the other dolphins in a school can determine which animal is producing the sound. By all appearances, a dolphin can determine quite exactly where a sound is coming from. Thus if a particular dolphin whistles a series of rapid, short, powerful signals, these might convey the following information: Sam is whistling, he is to your left, and he is very upset about something. Unlike Lilly and others, the Caldwells do not believe that dolphins have a particular SOS signal, but that they express themselves with their own individual whistle signature.

Whistling seems to be especially important in the relationship between cow and calf. As soon as the cow gives birth to her calf, she whistles almost constantly for several days. It is apparently during this period that the calf learns to recognize its mother's signature, a piece of learning that will stand it in good stead if it gets lost. The calf, too, often whistles during these first few days, and it is quite certain that the cow also learns to recognize the signature of the calf. The whistle of a calf was studied from birth to eighteen months and was found to remain the same.

The whistling of one animal in a school stimulates the others to do the same. It has been shown that response whistling does not begin until the first animal's whistling is completed. If two animals begin whistling simultaneously, one of the animals will often stop and wait until the other has finished.

The question of how and to what end the dolphin uses the pulsing sound that is not directly involved in echo-location seems to be considerably more complicated than the corresponding questions about whistling. In the first place, we cannot say for sure when the pulsing sound is being used for echo-location and when it is not. Nor do we know the limits of the animal's capacity to analyze sound. Another limiting factor in the study of dolphin vocalizations is that for positive identification of the sound-producing animal, the animal must be isolated. But isolation itself reduces the stimulus for vocalization, and we find ourselves in a vicious circle. As mentioned earlier, the Caldwells believe that the pulsing sounds to which no echo-locating function can be ascribed are used in emotional contexts. The following discussion will show that I believe it is this very portion of dolphin sound production that may be used in complex communication about things other than emotional relations.

In the same way that we discussed the number of informational bits we can receive by means of our hearing, we can also discuss the number of such bits the sound-generating apparatus can produce per unit of time. Since the dolphin employs frequencies 4½ times higher than the ones we use, its sound-producing apparatus can manage 4½ times as much information per unit of time. The dolphin appears to have two separate sets of sound-producing apparatus, one in each half of its blowhole. They can be used simultaneously. This means that the dolphin ought to be able to emit 9 times as much information as we can per unit of time. All according to Dr. Lilly in *The Mind of the Dolphin.*

Now the fact is that when used together, these two sets of sound apparatus produce a stereophonic effect, which ought logically to increase their informational capacity in relation to ours beyond a factor of 9. If we say that the dolphin can emit 10 times as much physical information via sound as we can, then that is probably an underestimate.

Linguists Are Skeptical

That other animals than human beings might have a language comparable to our own seems like an absurd idea to most people, even to most educated people. At a linguistics seminar held in Uppsala in the spring of 1969, innate linguistic capacity in human beings was discussed on the basis of a book by Eric Lenneberg, *Biological Foundations of Language.* The seminar outline read in part, "Language is species specific. A communications system comparable in quality and quantity to that of the human species does not exist in any animal. Nor can human language be taught to any animal."

I would like to argue that the intricate combination of phenomena necessary for a complex language unlike ours could very well exist in other animals, for example, dolphins. A nonhuman language of this kind, in order to be called a language in our sense of the word, must be able to convey information about the past, the present, and the future and be able to express something about the speaker, his plans, his problems. But it may employ a logic that is entirely foreign to us, and treat information in a way that seems to us backwards.

The same seminar produced a long list of inborn biological (morphological, physiological) prerequisites for human language. By and large, these prerequisites coincide with those mentioned before—a well-developed brain, something to talk about, and something to talk with. A comparison between dolphin and man in accordance with this list shows that on every point where we have adequate information, dolphins are superior to human beings. The points on which we lack definite information involve the brain's function. We believe we can say, however, that the dolphin has more cerebral cortex left over for higher processes than we do.

After having heard all this, the linguists were prepared to alter their introductory observation, "Language is species specific," to read, "Human language is species specific," which is a meaningless assertion. The contention that human language cannot be taught to other animals is also open to question. As a matter of fact, dolphins can imitate humans better than we can imitate them. No one has yet seriously attempted to teach dolphins to read by making use of their acoustical abilities in their natural element, water.

How Might a Dolphin Language Be Constructed?

How do we imagine the construction and function of a dolphin language? In *The Mind of the Dolphin,* John Lilly maintains that we must base our speculation on the fact that the dolphin lives its life in the sea. We know that dolphins acquire their view of the world primarily by means of sonar. When a dolphin examines a fish, it sends out a sound and gets back an echo. It is possible that for the purpose of describing this fish to another dolphin, the first dolphin sends out the echo he himself received. In any case the dolphin has the possibility, in theory, of using its vocal apparatus to imitate the so-called Doppler effect and thereby describe complicated movements. (The Doppler effect is the phenomenon in which the constant tone emitted by a train whistle as it passes a motionless person is heard by that person as a sound of continuously changing pitch.)

Nevertheless it seems unlikely that the dolphin can imitate its surroundings exactly, though it is possible that it can send simple sound pictures with the same degree of accuracy as a child's drawing.

It may be of interest to make a comparison with the communication methods of Stone Age man. When such a man wanted to impart something he had seen to another man, he had first to translate his inner picture into words, and the receiver of the message had then to translate the words into an inner picture, which is still in fact our commonest way of communicating. As an alternative, the speaker could make a drawing in the sand or the snow. In our modern society we have new means of communication, the most recent great achievement being television. What we can do with the help of TV is a thing that the dolphin may be able to do without artificial assistance, by quite simply retransmitting the echoes it receives from its environment. Such a communications system might in some ways be superior to ours. There would be fewer mistakes and misunderstandings, since the dolphin eliminates the intermediary steps. If this view of Dolphinese is correct, it may mean that the dolphin has less use of symbols than we. But it also means that the dolphin's communications are more dependent than ours are on computerlike capacities in memory and reproduction.

What Do Dolphins Communicate About?

Dr. Gregory Bateson has presented some interesting comparative views on what a dolphin language might be like. Even if we disagree with Dr. Bateson's guesses as to what dolphins talk about, his discussion of the problem is informative from the point of view of communications theory.

Like most ethologists, Dr. Bateson maintains that communication in mammals revolves primarily around the relations between different individuals. When the cat tries to tell its master that it wants milk, it meows. It has no word for milk. Instead, it emits the sound and executes the motions that a kitten makes for its mother in a similar situation. So if we attempt to translate the meowing, it means not "milk" but "mama." Or more accurately, "dependency, dependency." Thus the cat is talking about the relations between individuals, and it is then the business of the recipient to figure out that milk is what it wants. This need of a deductive step represents the difference between the communication of preverbal mammals and our language (and that of bees!). The big new feature in the development of human language was the possibility of specific expression about something other than the relationship between individuals.

The question at hand is whether dolphin

communication amounts merely to a complex communication about the relations among the different animals—of which they certainly have a need considering they are very social and are equipped with wicked teeth—or whether dolphins are also capable of detailed communication about other things. Bateson subscribes to the first alternative.

Human beings communicate more about their relations to each other than about anything else—often unconsciously. Most of this communication takes place with the help of movements and so-called paralinguistic sound signals —body movements, involuntary contraction of muscles usually controlled by the will, changes in facial expression, hesitations, alterations in the tempo of speech and movement, vocal overtones, irregularities in breathing, and so forth.

If we want to know the meaning of a dog's bark we must look at its lips, at the hair on the back of its neck, at its ears, tail, and throat. These expressive parts of a dog's anatomy will tell us what it is he is barking at and what his relationship to that object is, as well as what he will do during the next few seconds.

In all mammals, the sense organs are of primary importance in conveying messages about relationships. We look at the dog's eyes, nose, and ears. We cannot know much about communication between dolphins until we know what one dolphin can read from another's use of sonar—direction, volume, and pitch.

According to Dr. Bateson, dolphins have, in

the course of their adaptation to aquatic life, lost their ability to express themselves with their faces. They have no outer ears to lift or lower, little or no hair to raise. The vertebrae in the neck of many species are fused into a block and the body is impersonally streamlined, so that the expressiveness of its various parts has been lost. Bateson contends that in dolphins, sound production has taken over the communicative function that other mammals accomplish with facial muscles, wagging tails, clenched fists, widened nostrils, and the like. In terms of communication, the dolphin is the opposite of the giraffe—it has a voice but no neck.

Consequently it is easy to imagine that in communicating about the relations between individuals, the dolphin has replaced body movements of various kinds with paralinguistic signals (which are auditory by definition) corresponding to our laughter, our grunts, sobs, and varied respiration. We recognize such paralinguistic signals in other mammals even if our guesses as to their meaning may be wrong. But the underwater vocalizations of the dolphin resemble none of these, and it is hard to imagine that all the pulsing sounds they make in addition to echo-locating are nothing more than paralinguistic signals. Here Bateson is of the opinion that they do indeed use sound to communicate about their relationships, but that in so doing it is not merely paralinguistic signals they employ.

Digital and Analogic Communication

For the purposes of the ensuing discussion we must introduce two new concepts—digital and analogic communication. Language is digital, and paralinguistic signals are analogic. Digital communication employs a number of accepted symbols—1, 2, 3, x, y, and so forth—which mean different things when placed in different positions. The symbols themselves have no simple correspondence to what they stand for. The numeral 5 is no larger than the numeral 3. The numeral 5 is the *name* of a quantity. It is nonsense to speak of someone's telephone number as being larger than someone else's, because the telephone exchange is a purely digital computer. It is not fed with quantities but with names in different positions.

Analogic communication, on the other hand, makes use of quantities that correspond to the real quantities in question. If an English-speaking mathematician is required to read a paper by a Japanese colleague in Japanese, he will not understand anything of the text, but he will be able to understand at least some of the mathematical curves—which correspond to real quantities. The curves are analogic, while the Japanese text is digital.

Spoken language is almost purely digital. The word "large" is no larger than the word "small." Paralinguistic signals are analogic. The sound level, the length of the pauses, have a relation to the size of the message to be conveyed.

Bateson now puts forward the hypothesis that dolphin vocalizations are digital expressions of the relationships between individuals. Man has a number of words for expressing these relationships, words such as love, respect, dependency, and so forth. But these words function poorly in the actual discussion of the relations between individuals. A girl pays more attention to a boy's movements and paralinguistic signals than to the words "I love you." According to Bateson, dolphins are dependent on "words" for expressing their feelings.

Bateson's hypothesis is very tantalizing, and attests to the imagination and creativity of its author. But there is one serious objection, which no doubt the reader has already raised. Dolphins can presumably "hear" one another's intestines, and in that way have access to an analogic communication system for describing relationships between individuals that would appear to be superior to ours. We will probably have to suppose that dolphins' digital language, if indeed they possess one, exists for the treatment of subjects beyond those touching on the interrelationships of individuals.

Complex Language in Dolphins Is Still Unproven

Let us make it perfectly clear, however, that no one has yet demonstrated the existence of a complex language in dolphins—though the day cannot be far off when we will know for sure. Suppose we repeated the communication experiment described in the section entitled "Intelligent Behavior in Dolphins."

Instead of showing the female different kinds of lights, we could present her, under water, with various objects with which she was well acquainted—for instance a stone, an underwater plant, and a fish—each one displayed beside a key. Similar objects and keys would be presented to the male. The female would be trained to tell the male to depress the key next to the object we pointed out to her. We would then register the echo received by the female the first time she "listened" to an object and then compare this with the sound she sent to the male when she communicated. By repeating this kind of experiment with different objects, different varieties of the same kind of object, animals with different levels of experience, and so forth, it ought to be possible to gain some knowledge of the dolphin's vocal language, if such a language does in fact exist. We could answer questions such as: Does the dolphin use symbols? Is the initial description of a stone dependent on the stone's shape or size? Is there a word for stone? How does a dolphin describe a real fish, as opposed to an iron fish of the same size and shape? After long and arduous experimentation it should be possible to test our ability to understand the dolphin language by imitating their signals with sound-producing equipment.

These experiments would doubtless be difficult and expensive, and in particular would test the patience of the scientists involved. Tactically, it might be wise to first give the dolphins a chance to meet us halfway, by trying to teach them a human language.

We Must Take the First Step Toward Communication

A dolphin is not likely to have any trouble learning to recognize a letter shown it under water. It is also very probable that the dolphin has a large acoustical memory and could readily learn to recognize the word "dolphin" and to distinguish it from the word "man." By showing dolphins objects that are well known to them and at the same time showing them our words for the same objects, it ought to be possible to teach them quite a number of words. People have succeeded in teaching two- to three-year-old human children to read. The principle is to proceed from the whole to the parts. The same principle applied to dolphins in conjunction with entire words under water might perhaps yield results even more astonishing than those achieved with human children.

At all events we must assume that *we* will have to take the first step toward eventual communication with dolphins. Any successful attempt will have to be based on an acceptance of the dolphin's natural life in the sea. We must begin by placing our symbols or objects beneath the surface of the water. Sounds in air are too difficult for dolphins to grasp, just as sounds in water are difficult for us. Dolphins cannot hear such signals. The dolphin's visual cortex is not the part of its brain we are principally interested in. The visual part of its brain has probably not developed appreciably during the 50 million years dolphins are presumed to have been aquatic. We must concentrate on their acoustical abilities.

Attempts at Communication with Chimpanzees

In our attempts to learn to communicate with apes, or rather to teach apes to communicate with us, we have recently had to learn a lesson that supports the truth of what we have just been saying. A number of experts have tried to teach apes our spoken language. People have gone so far as to raise apes along with their own children. Not even in these experiments, which lasted for several years, has anyone succeeded in getting an ape to do more than repeat a few words. During the last few years, however, systematic studies were made of the chimpanzee's spontaneous gestures, and it was discovered that chimpanzees and children use sign languages that are similar in many ways. This observation prompted Dr. W. N. Kellogg to attempt to teach a chimpanzee the sign language of the deaf and dumb. In sixteen months his chimpanzee learned to understand and use 19 signs. Most of these belonged to Standard American Sign Language, but a few were original chimpanzee. Some of the signs were verbs and adjectives that the chimpanzee could use correctly in varying contexts. In the course of these sixteen months, the chimpanzee also learned to understand many more signs than it could use. These results outshine anything previously accomplished in the way of human-ape communications. We have met the chimpanzee halfway. We ought to be able to accommodate the dolphin by lowering our symbols into the water.

The Learning Potential of Other Large-Brained Animals Is Still Untested

Displaying letters of our alphabet to dolphins under water must not be seen merely as an attempt to teach dolphins to read our language. It should be placed into a larger context. It is an attempt to investigate the learning potential of the dolphin brain with the aid of some of the learning sets we have successfully used on ourselves during the last 5,000 years. These learning sets are based on our own language, which is, to an unknown extent, innate, and which is therefore much less well adapted to the potential talents of the dolphin than to our own. If we fail in this attempt, we will have to use our imagination to invent and manufacture other learning sets that better suit the dolphin.

We must learn to approach the dolphin properly, by means of its hearing. By the same token, we have every reason to reconsider our approach to the elephant. With a certain amount of success we have taught elephants to distinguish between visual symbols and have even taught them to write a little on a machine with easily legible letters. The elephant's vision, however, is very poor, while the sense of touch at the tip of its trunk is said to be fantastic. Why not try braille?

We must make a serious attempt to explore the functional mental potential of large-brained animals in general. And all the while we must bear in mind the past 5,000 years of our own history, during which we have managed to learn so much, both for better and for worse.

Who Was Moby Dick?

Moby Dick was the white sperm whale who proved to be more than a match for the whalers in Herman Melville's famous novel from the days of the great whale slaughter in the middle of the nineteenth century. We hope that this chapter on the large-brained animals will have aroused the reader's interest in the enigmatic creature that possesses the largest cerebral cortex on this earth. The following description of the sperm whale is presented with its large brain constantly in mind. Is there anything in what we know of the sperm whale today that gives us an indication of the purpose to which this enormous animal puts its cerebral cortex?

PART II

The Sperm Whale–
A Large Question Mark

CHAPTER THREE

A Survey of the Whales

For purposes of classification, zoologists divide the order of whales (Cetacea) into three suborders: the extinct fossil whales (Archaeoceti), the present-day toothed whales (Odontoceti), and the baleen whales (Mysticeti). Zoologists are agreed that all whales are descended from mammals that lived on land, probably otterlike, fish-eating predators that evolved steadily toward a more and more complete adaptation to aquatic life. Of course we do not know why these ancestors of the whales evolved in this direction, but on the whole the sea offers an environment that is richer in nourishment than the land. And it is not improbable that the special characteristics of mammals—respiration of air,

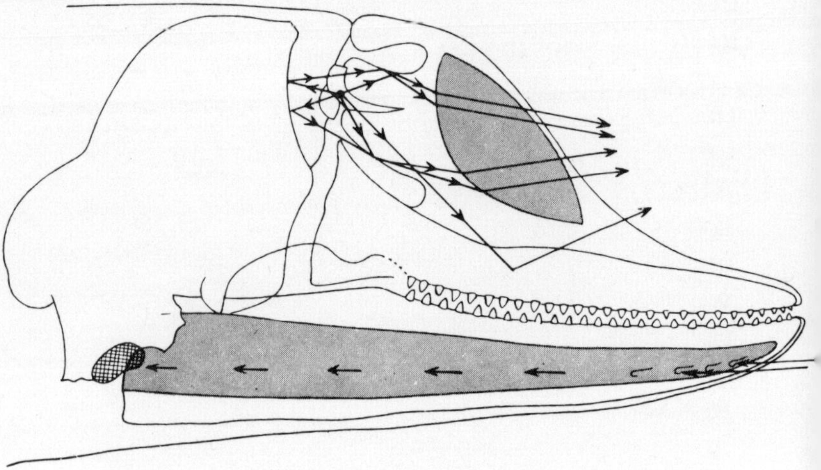

A sound produced in the nasal cavity is broken and focused by the fat body in the whale's forehead, and thus emitted as a beam. Incoming sounds are led to the inner ear through the fat body in the lower jaw.

high body temperature, development of the young inside the mother—were advantageous in the new environment and permitted these new aquatic creatures to compete successfully with the old. For that matter, it is not only mammals that have recolonized the sea. We find examples of this phenomenon in many other groups of animals such as penguins, marine iguanas, sea snakes, and sea turtles. It is perhaps more interesting to speculate on why the different aquatic mammals—whales, seals, otters, beavers—have evolved in such different ways. It is evident that the basic plan of mammalian body construction can be turned to account in water in many ways, even if this basic plan varies greatly from one group of mammals to another.

Specialists seem to be agreed on the fact that the primitive, extinct whales, whose fossil skeletons have been found principally in North Africa, are not the direct ancestors of the whales living today, but that the three suborders developed from a common origin. The fossil whales were well armed with teeth, and we suppose that they lived about like the toothed whales of our day. It is interesting to note that they do not seem to have had anything corresponding to the fatty accumulations in the heads of present-day toothed whales, nor to any of the other structures that assist or are thought to assist in echo-location. Nor did they have specialized teeth or baleen. Also, their brains seem to have been smaller, all of which indicates that they had not developed the systems that present-day whales use for locating their prey by echo-sounding, but that they hunted fish with the help of their eyes roughly the way crocodiles do. What caused them to die out is unknown, of course, but it is possible that the suborders alive today outdid them in the contest for survival.

The second suborder consists of the toothed whales, and these are the ones that interest us most at the moment. The best known of these are the dolphin, the porpoise, and the sperm whale, but there are quite a large number of different families and species. Many of these species are so unfamiliar to us that we know of them at all only through some few stranded specimens. Despite the fact that no one has ever seen them alive, they must nevertheless exist in sufficient numbers for the populations to survive in a natural manner. These whales are not scarce because man hunts them, but only because we know so amazingly little about the many different types of life in the sea. Rare species of this kind occur particularly in the family of beaked whales (Ziphiidae), where the genus *Mesoplodon* includes at least nine species known, in most cases, from fewer than ten individual specimens. This family is interesting because its members are equipped with two large, peculiar teeth in the lower jaw. It has been suggested that these species live exclusively on the arms of squid, which they tear from the body. But the teeth are found only in the male, which probably indicates that they are a sexual characteristic of the same type as the narwhal's tusk (see below).

A fact not generally known is that there are several different species of toothed whales that live entirely or partially in rivers. The species that stick to fresh water are primarily those within the family Platanistidae. For various reasons, it seems probable that this has been a characteristic feature of theirs from the beginning, that is, that they did not secondarily return to fresh water from the sea. On the other hand, this *is* the case with the fresh-water forms of other whale families—for example the Irrawaddy dolphin (*Orcaella brevirostris*), which is one of the Delphinidae, and lives in the Irrawaddy River in Burma. It is typical of the Platanistidae to have

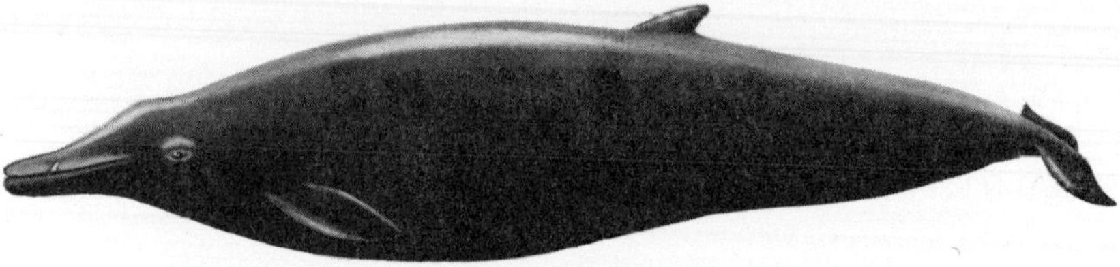

Sowerby's whale *(Mesoplodon bidens)*

Amazon River dolphin *(Inea geoffrensis)*

a relatively long, mobile neck and degenerate eyes. Species are found in most tropical river systems: the Ganges River dolphin *(Platanista gangetica)*, of the Indus and Ganges rivers; the Chinese river dolphin *(Lipotes vexillifer)*, of Tung Ting Lake in China; the buoto, or Amazon River dolphin *(Inia geoffrensis)*, of the Amazon and Orinoco rivers, where we also find the La Plata dolphin, *(Stenodelphis blainvillii)*. Unfortunately we know very little about these various fresh-water whales, because the rivers in which they live are turbid and it is difficult to follow the animals for any great distance in order to observe them.

The Platanistidae are usually counted as the first family of toothed whales, and are followed by the previously mentioned beaked whales (Ziphiidae). The beaked whales include a number of rare species (at least from the human point of view), but one of the few species that is better known is the bottlenose whale *(Hyperoodon ampullatus).* Like other beaked whales it feeds on squid and seems to prefer the smaller varieties. More than 10,000 squid beaks have been found in the stomach of a single specimen. It is also known for being a skillful and persistent diver, reported to have reached depths of over four thousand feet when harpooned. It lives in small schools where the old males, by butting and striking with their flukes, fight fiercely for power and females. The young, born in the spring, are nearly 10 feet long, although adults reach a length of only 23 to 30 feet.

The third family of toothed whales is the sperm whales (Physeteridae), of which the best-known member is the sperm whale itself *(Physeter catodon)*, to whom we will return. Less well known is the fact that there is also a smaller, similar species, the pygmy sperm whale *(Kogia breviceps)*, which grows to only 10 or 12 feet in length. Unfortunately, we know it by and large only from stranded carcasses, which is the case of so many species of whale. The sperm whales are followed by the family Monodontidae, which includes the white whale and the narwhal. The white whale, or beluga *(Delphinapterus leucas)*, is an arctic species that reaches a length of 10 to 16 feet and, as the name implies, is white in color. It is interesting in that it seems to be unusually skillful at producing sounds, and has even been called the "canary of the sea" on account of its ability to chirp, whistle, and warble. Various studies have fixed its auditory range in the area between 150 and 153,000 Hz. (Our own range of hearing is roughly from 16 to 20,000 Hz.) We do not yet know if the white whale is actually more proficient than other species at producing sound, or if it is only that its signals are easier to hear. Observers have thought they could distinguish various calls and warning cries. The white whale too lives in small schools that may join together into larger herds during migrations. The period of gestation is about ten months and the young, about 4 feet long, nurse for an entire year.

The narwhal *(Monodon monoceros)* has always aroused curiosity because of the long, spirally twisted tusk that the male carries in his upper jaw, which was the model for the horn of the mythological unicorn. The most likely explanation of this strange tusk is, of course, that it is a sexual characteristic of the same type as a buck's antlers and a peacock's tail. Most of the odd and "unnecessary" forms, colors, and even behavior patterns in the animal world have, in the light of recent research, proved to be sexually determined signals. Perhaps the males use their tusks in some sort of pushing contest, since narwhals with broken tusks have been found with another tusk plugged into the pulp cavity. The size of the tusk might be explained by such fights, and perhaps also by a sexual selection where the female chooses the male with the largest tusk. Of course, the tusk might also function as a tool in obtaining food. But in that case we might ask ourselves why the females do not require such a tool. In the end we will have to be content with the fact that the riddle is still unsolved. The narwhal is a typical arctic animal and an important quarry for the Eskimos, who eat its skin raw. It contains over 30 milligrams of vitamin C per 100 grams of weight and is their most important source of this vitamin, which, after all, they cannot obtain from vegetables.

The next family of whales is the dolphins (Delphinidae). The best-known species is the common dolphin *(Delphinus delphis)*, which is found in all warm seas. It reaches a length of about 7 feet. This is the species that most often follows ships, since it can easily manage speeds up to 20 knots. It is not this species, however, that is kept in aquariums, but rather the bottlenose dolphin *(Tursiops truncatus)*, which, thanks to the fact that it seems to thrive in captivity, has become the best known of all whales. The majority of the studies and results quoted elsewhere in this book are based on experiments done on the bottlenose dolphin, and most popular books about dolphins actually treat only this one species. This is not to say that there is any reason to believe the bottlenose dolphin differs significantly from other toothed whales, and most of the results of research done on bottlenose dolphins in regard to echo-location, communication, and so forth can no doubt be applied to other species as well. Since the bottlenose dolphin is so comfortable in captivity that it even reproduces, we have been able to follow its social life at close hand. We have been able to observe and film the birth of the young (which in whales, as opposed to what is normal for other mammals, takes place tail first). We have seen nursing, growth, and such motherly cares as lifting the calf to the surface to breathe, particularly if it is in difficulty. This act of lifting to the surface also occurs among adult dolphins, who, like elephants, take care of and directly support an injured member of the herd. Dolphin females have even been observed carrying calves that have been dead for weeks, even when only the head of the body remained.

Various popular books are filled with information on the bottlenose dolphin's great capacity for learning circus tricks. For the moment we will only remind the reader that the bottlenose dolphin has succeeded in learning a thing no other animal, as far as we know, has ever been able to manage: namely, to perform *new* tricks, where they were rewarded every time they themselves invented some new variation in their jump, or the like, which shows that they had somehow grasped the idea of "novelty" in this context.

In the same family as the bottlenose dolphin we find the killer whale *(Orcinus orca)*, much discussed because of its alleged bloodthirstiness.

Common dolphin *(Delphinus delphi)*

Bottlenose dolphin *(Tursiops truncatus)*

Killer whale *(Orcinus orca)*

Common or harbor porpoise *(Phocoena phocoena)*

It has even been accorded the doubtful honor—along with the wolf in the Soviet Union and the tiger in China—of being subjected to the machinery of modern human warfare. In the summer of 1955, the American air force dropped depth charges on several thousand killer whales off the coast of Iceland after they were accused of destroying fishermen's nets. As is nearly always the case with large predators, recent research has shown that their reputation for viciousness is much exaggerated, and that the mischief for which they have been blamed was, in most cases, ultimately the result of human

Even the so-called killer whale, which is known for its healthy appetite for all manner of animal protein, is, in captivity, friendly toward human beings, whom it can hardly regard in a positive light.

actions. Even if there are many reliable reports of killer whales attacking, say, a larger whale and ripping out its tongue, and even if the killer whale, like the shark, seems to fall victim to a kind of herd frenzy in the presence of wounded prey, nevertheless their relationship to the larger whales seems to be about the same as that found between large terrestrial predators and their prey. That is to say, killer whales constitute no danger to a healthy adult whale, which can easily defend itself with blows of its tail. Sick and old individuals, on the other hand, as well as females with newborn young, can get into trouble. If killer whales were as bloodthirsty as the more imaginative reports suggest, it is hard to see how their helpless prey have managed to survive this long. For that matter, a killer whale held in captivity in an American aquarium has turned into a veritable housepet, who relishes having his teeth brushed before the public.

Another well-known species in this same family is the pilot whale *(Globicephala melaena)*. Their propensity for falling into mass panic permits them to be driven like cattle either into narrow fjords, as in Norway, or straight up on land, as in the Faroe Islands. Pilot whale hunts on the Faroe Islands are well known from an endless number of descriptions in words and pictures, and the spectacle of the whales being driven in and slowly stabbed, cut, and beaten to death with every imaginable available tool has become a rewarding source of snapshots for many tourists. Since the hunt can

The result of a pilot whale slaughter on the Faroe Islands. The whales are divided among the participants by means of symbols cut into their skin. In the rush to lay claim to as much of the catch as possible, this is often done while the whales are still alive.

no longer be defended on economic grounds, it has been decided to preserve the bloodbath for its entertainment value. Traditionally, even divine services are interrupted if a pilot whale hunt is in the offing, a fact that might give us pause for reflection. Naturally the pilot whale has become fairly rare, and the Norwegian practice of confining a school in a fjord and butchering them one by one as needed over the course of several weeks—until they grow too emaciated to be of any use—has had to be abandoned because the herds have disappeared.

Mass strandings also occur without the help of man, and in other species than the pilot whale. In 1950, 97 whales were stranded on the Orkney Islands and 167 in Scotland, and 1,400 animals are supposed to have been stranded on Cape Cod in 1874. These are probably cases of animals being pursued and, like cattle, blindly following the herd. The impulse to do this is said to be so strong that when stranded animals are towed back out to sea, they return to the beach. Among the other species that fall victim to mass strandings, the false killer whale *(Pseudorca crassidens)* is the best known. There are at least twelve reports of the phenomenon in this species.

The next family, Phocoenidae, includes the common, or harbor, porpoise *(Phocoena phocoena)*, which lives in coastal waters around the North Atlantic and even enters the Baltic and the Black Sea. About 5 feet in length, it is the smallest of all whales. But its size has not kept it from being hunted, particularly in the Little Belt, along the east coast of Jutland where before the turn of the century up to 2,000 animals were taken each year. This harvest quickly reduced the species to unprofitable proportions. The hunt was based on the migration the porpoise makes every spring and fall between the Baltic and the North Sea, although a large number of porpoises remain in the Baltic over the winter, which can be fatal in years of heavy ice. In spite of the fact that porpoises are regularly seen along the coast, we know as little about them as we know about most other whales. This species is being studied by Swedish ethologists, and attempts have been made to repeat the communication experiment previously described with dolphins, so far with little success. But the behavior of this porpoise is interesting, showing rank order as well as aggression, the existence of which in small whales is sometimes denied.

The last suborder of whales is the one that perhaps most people associate with the concept "whale," namely, the baleen whales (Mysticeti). What distinguishes them from the toothed whales is the baleen or whalebone that hangs from the palate in long folds and which developed from the faint ridges the reader can feel with his tongue on the roof of his own mouth. Teeth begin to develop in these whales too, but disappear again in the embryonic stage. Baleen was a very lucky "invention" since it made it possible for its owner to move from a higher to a lower level of the food chain (where ten times the food was available). With the help of its baleen, the whale can feed directly on plankton instead of eating the animals that eat plankton. In this way the baleen whales correspond to grass-eating animals on land, since their baleen permits them to graze in peace and quiet on the "meadows" of the sea. To be sure, they eat a kind of plankton that is already one step removed from the plants at the very bottom of the food chain, namely various kinds of small crustaceans, but even these small animals exist in such enormous numbers that among baleen whales we find the largest animals that have ever lived. The baleen whale's method of feeding is very simple: it finds a shoal of krill (a small crustacean about two inches in length), opens its gigantic mouth, and engulfs the shoal along with several tons of water. The water is then filtered out through the baleen, which is fringed on the side facing the tongue, and the krill catch on this fringed wall. The whale then uses its enormous tongue, which can have a surface area as great as that of a small room, to lick this plankton porridge back from the baleen and down its throat. Since the mouth has to be very elastic, most baleen whales have marked folds on their "throats," which can occupy as much as a third of the length of the entire body.

The baleen whales are divided into three families: the gray whales (Eschrichtidae), with only one species, the gray whale *(Eschrichtius glaucus)*; the rorquals (Balaenopteridae); and the right whales (Balaenidae). The gray whale is noteworthy in that it retains several primitive characteristics, such as a fairly rich growth of hair and relatively undeveloped baleen. In addition to which the gray whale, as opposed to its larger relatives, can to some extent survive on land without being crushed by its own weight. It seeks out shallow water for propagation, and the female gives birth to her young in sheltered lagoons. This may be a primitive characteristic, and it unfortunately led at one time to a reduction of the population from about 40,000 animals to the brink of extinction, since it was easy to kill the females in coastal lagoons, particularly just after parturition. Amazingly enough, however, the gray whale was declared protected in the 1930s, and since then its numbers are thought to have increased to about 3,000 individuals, despite poaching. This has naturally led to efforts to have the protection withdrawn, even if so far no one has proposed reviving the profitable practice of hunting new mothers.

The rorquals include, first of all, the famous blue whale *(Sibbaldus musculus)*, whose record length of 112½ feet makes it the largest animal on earth. This becomes even more impressive when we compare its 187 tons with the largest of the dinosaurs, which had to be content with 38 tons. The blue whale still occurs in all the oceans, though it is now seriously threatened by extinction, and undertakes what can be very long journeys in search of whatever pastures have the highest yields at the moment. It usually lives in small schools consisting of a male and one or two females, with their young. But we do not know a great many details, since a school of blue whales is hardly given the time to display any of its natural behavior between the moment it is sighted and the moment the guts of its members are torn to shreds by the tips of exploding harpoons. However, breeding occurs during the winter, the gestation period is about ten and one-half months, and the young are already 20 to 23 feet long at birth. They then grow at a prodigious speed, increasing at times by as much as 200 pounds *per day* in permanent body weight. By the end of seven months, the calf is over 50 feet in length, and after two years it is 70 to 75 feet long and, astonishingly enough, sexually mature. We can only speculate as to the causes of this remarkable adaptation, but it is probable that size is the blue whale's principal protection against various enemies, and that it therefore has to attain this size relatively quickly. Rapid rates of growth are typical of all whales but particularly conspicuous among the largest species.

The blue whale swims quite slowly, since no speed is required for gathering plankton. Nor is it known for being an especially outstanding diver, another ability not needed by a species that feeds near the surface. On being pursued, however, it can reach speeds approaching 20 knots, and can stay under water over half an hour if necessary.

Two other species very similar to the blue

whale are the fin whale *(Balaenoptera physalus)* and the sei whale *(Balaenoptera borealis).* They are smaller, however, reaching maximum lengths of 82 and 62 feet respectively, and also differ somewhat in their choice of food, since both of them eat small fish in addition to plankton. Both species are fairly common off the coast of Norway and the west coast of Sweden, or rather were fairly common. The Norwegians succeeded in decimating particularly the fin whale and reducing its numbers substantially up until 1957, when it had become so rare that such whaling was no longer profitable. The sei whale holds the known whale speed record of 30 knots (almost 35 mph), but since it is no better than other whales in perceiving that a whaling ship is dangerous, this speed does not help it much. Ever since canned dog food became popular, there has been a growing interest in baleen whales, including the minke whale or piked whale or lesser rorqual *(Balaenoptera acutorostrata)*, which only gets to be a little over 30 feet long. Baleen whaling is conducted on Iceland, for example, where such whales are still to be found.

Another well-known rorqual is the humpback whale *(Megaptera novaeangliae).* It has wartlike knobs on its head and on the edges of its unusually long flippers, and gets its name from the small, humplike fin on its back. Like the other rorquals it is found in all the oceans, and is particularly well known for its conspicuous mating behavior, in the course of which the animals leap entirely out of the water and slap each other playfully with their flippers, which makes a sound like artillery fire. The humpback whale is of special interest to zoologists because of the various parasites it harbors on its fifty-foot body. Several of these, including a species of goose barnacle (a relative of the common acorn barnacle), are found on the humpback whale and nowhere else. Like other whales, they also play host to the common whale louse, to marine worms, boring lamellibranchs, sea spiders, and a lot of other things. The humpback whale has come into the limelight recently as a result of the discovery of its capacity for communication, which we will discuss in greater detail further on.

The final family of whales is the right whales, so called because they were the only whales hunted during the early years of whaling. The rorquals have an unpleasant habit of sinking when harpooned, for which whalers can now compensate by blowing compressed air into the (usually) dead animal, but before the invention of this device, only the right whales were right for whaling. It is also right whales that are most often pictured on medieval maps, since the black right whale *(Eubalaena glacialis)* used to be common off the coasts of Spain and France. Industrious hunting eventually made it so rare that it was declared protected, a legality one could always afford when no animals were left.

As with the Greenland right whale *(Balaena mysticetus),* now also nearly extinct, it was primarily the baleen or whalebone for which there was a market, at least in recent times. Not only do right whales have the longest whalebone plates (up to 13 feet), but their whalebone is also the strongest and most flexible. These qualities made this whalebone much in demand for use as an elastic element, particularly in the type of corset that seems to have been so popular in turn-of-the-century cabarets. And even if it was possible to take an entire ton of whalebone from a single whale, the market was apparently insatiable. Female vanity became the ruin of the whale, the same way that the fashion in plumes was once about to exterminate the ostrich, and that the popularity of fur coats today will end in the extinction of the ocelot, the leopard, and the other spotted cats unless some action is taken.

That the right whales are so splendidly equipped with baleen is probably due to a scantier supply of plankton in their home waters, which means that they must strain more water per unit of food. Of course we know very little about these species, but both occur in schools, and the Greenland right whale is said to be a capable diver, at least after it has been harpooned.

Whales demonstrate various methods of providing for themselves, and this variety is naturally reflected in their behavior and in the quality of their mental equipment. Nevertheless, there is reason to believe that most of the discoveries made so far on individual species are applicable to all or to most types of whales. Echolocation must exist in all whales to a greater or lesser degree, even if it has not yet been observed. And even if we know of specific long-distance communication only in the humpback whale, it would be astounding if this system were not used by all species. We might ask ourselves how else they find each other. We have observed intelligent social behavior in only a few species so far, but it must exist in others as well, even if the baleen whales probably have no use for any great intellectual capacity, considering their way of life.

Any survey of the whales must conclude with the sorrowful confession that we know amazingly little about them, in spite of the fact that they are, or at least have been, of such tremendous importance for us. We can only hope that this trend will now be broken. For one thing, increased knowledge about an animal can make it possible to protect it. For another, such knowledge might arouse an interest in the animal and encourage people to oppose its economic exploitation. In one way, knowledge has already ignited one spark of hope: if it is true that whales can maintain contact with each other over very great distances, then it is possible that they can survive even after their numbers have gone down so far as to make whaling unprofitable. The great danger otherwise is in reaching the critical limit for population size, as happened with the North American passenger pigeon, once slaughtered in the millions for use in producing soap and candles and now completely extinct. But despite this hope, no one yet knows whether the whales or the whalers will reach this limit first.

CHAPTER FOUR

Introduction to the Sperm Whale

The sperm whale, or cachalot, is, as we said, a toothed whale, in fact the largest species in this suborder. Maximum length for the male has been reported as 83½ feet, while females are considerably smaller, seldom more than 40 feet. At the present time, however, adult animals generally reach lengths of about 60 and 32 feet, respectively. The appearance of the sperm whale is so characteristic that even in very old drawings one need never hesitate as to which whale is intended. The fatty accumulation, or more precisely, the oil-filled cavity, that beaked whales have in their foreheads has become so enlarged in the sperm whale that it completely covers the "beak" and makes up a third of the

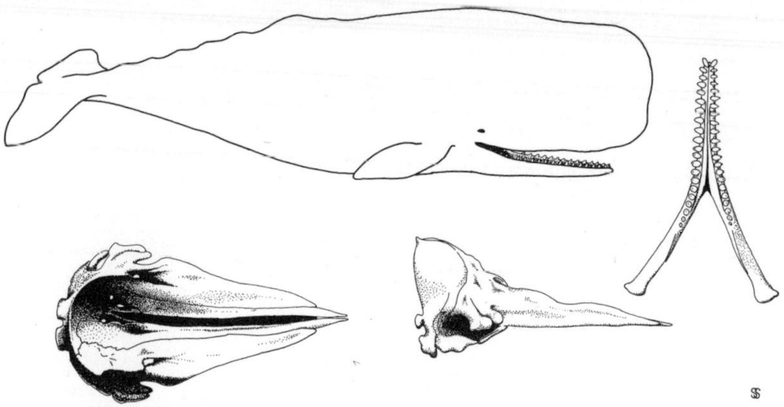

The sperm whale is distinguished from other whales by the enormous oil receptacle in its forehead, which has brought about great changes in the shape of the skull. The picture shows the skull from above (left) and from the side. The lower jaw (from above) is also specialized.

body. There has of course been a great deal of discussion as to what might lie behind this remarkable formation. The most popular explanation seems to have been that it was some sort of buoyant body, a swimming bladder supposed to help the animal hold itself upright. Unfortunately, this hypothesis does not explain why the sperm whale should require such an organ when other whales get along without it. Since we now know that the fatty accumulation in beaked whales has a sort of sound-collecting and perhaps also a sound-projecting function comparable to the function of a radar screen on a ship, we can be fairly certain that the fat-body in the sperm whale is used in the same way. The size can also be explained, at least hypothetically. If we begin with the assumption that the fat-body is used in echo-sounding, then clearly its size bears a definite relationship to the frequency and strength of the sound waves the whale is using, and these in turn are adapted to the size of the prey. Thus the sperm whale uses its oil-filled head the same way a ship or an airplane uses its radar screen—to take bearings on various objects, in this case primarily the larger squid. The oil or fat is light yellow in color and of low viscosity. It is also called spermaceti (whence the name "sperm whale") and was, especially in former times, a coveted commodity for whalers, who simply cut holes in the heads of captured sperm whales and drew up the oil in pails as if from a well. In Melville's *Moby Dick* there is even a description of how a whaler falls into this hole and comes close to drowning inside the whale's head, which tells us something about the size of the cavity. Another measure of its size is that as much as five tons of spermaceti could be ladled from the head of a large male.

This enormous oil container naturally entails a number of other specialized characteristics in the sperm whale. The blowhole is not, as in other whales, on top of the head where it would be too far back, but at the front upper edge of the snout. This means that the sperm whale blows forward at an angle, which makes its fifty-foot spout of vapor easy to recognize. As in other toothed whales, the entire cranium is asymmetrical, that is, different on the right and left sides.

This is particularly true of the nose bones, of which the right one has degenerated entirely. The left eye is also supposed to be smaller than the right, which led whalers before the days of the harpoon cannon to sneak up on the animals from the left side.

The lower jaw has an interesting design. It is remarkably narrow, only 18 inches even in large animals, and equipped with a varying number of teeth (roughly forty), each about the size of a fist. Oddly enough, these teeth do not seem to be used primarily for holding the whale's prey. They do not even break through the gums until the animals reaches sexual maturity. Moreover, many animals have been found with teeth that were worn down or worn away, but they could obviously provide themselves with food. Instead, it seems that the teeth are used in battles between the adult males. Such battles have often been observed, and injuries to the teeth and even broken jaws have been noted in connection with them. The teeth lack enamel, and are used industrially in the same way as ivory—for buttons, piano keys, markers in games, and so forth. Another characteristic of the lower jaw is that it can swing out at right angles to the body, not only straight down but also to the sides. Despite this great mobility, it is well equipped with muscles, since there are many reliable accounts of whaleboats being bitten in two by harpooned whales. It should be added, finally, that we often find sperm whales with lower jaws that are deformed, bent in various directions, even twisted into a spiral, without their owners being emaciated or underdeveloped in any way. By examining the contents of their stomachs, we know that sperm whales usually consume their prey whole, so the appearance of the lower jaw means very little as long as the catch can be held long enough to be swallowed.

Apart from its head, the sperm whale does not differ greatly in external appearance from other whales. The pectoral fins are unusually short, a mere 3 to 6 feet, which we can compare with the roughly equal-sized humpback whale whose pectoral fins reach a length of from 15 to 20 feet. A dorsal fin in the strict sense is lacking. There is instead one large and several small finlike bumps on the upper part of the tail. The flukes are of course horizontal, as they are in all whales, which probably has something to do with diving and the fact that whales breathe air (although the air-breathing fossil ichthyosaur had vertical tail fins!). The flukes are large and well developed. Skin color is black to gray, the belly often lighter. As a curiosity, we might add that animals have actually been observed that were completely white, counterparts to Moby Dick. The skin is not completely smooth but furnished here and there with wrinkles and folds. These formations, along with the dorsal fin and other outgrowths, seem to exist in order to fill vacuums, or rather, spaces around the body of the animal where water would otherwise form eddies from low pressure. It might be mentioned that the fins of whales, with the exception of the pectoral fins or flippers, are completely lacking in bone as a structural element. Instead they are made of connective tissue from the skin. Otherwise, the skin is not as thick as one might expect, and the horny outermost layer is quite thin. It seems to be the 8-to-12-inch-thick layer of blubber that serves as a protection against injuries and enemies. The skin, finally, is very sensitive, particularly around the mouth, and as with other whales, it is a cherished home for many parasites.

The most interesting aspects of the whale body, which in its fundamental features does not differ so greatly from the body of any other mammal, are the special adaptations to aquatic life. These include such phenomena as the reduction of the hind legs, where by now only traces of the pelvis remain. External streamlining has involved other skeletal alterations such as the shortening of the seven cervical vertebrae into an almost rigid pillar of thin discs, and certain changes in the formation of the skull, as mentioned earlier. Streamlining has also led to complete retractability of the sexual organs into pockets. As we would expect, the greatest changes in the body's internal organs are found in the respiratory and circulatory systems. The lungs are not excessively large, as we might have supposed. On the other hand, respiration is much more efficient than in human beings, for example. Nearly 80 percent of lung capacity is replaced at each breath (we replace about 20 percent) and no less than 12 percent of the 16 percent oxygen in the inspired air is utilized, as opposed to 5 percent in land mammals. This effectiveness of the respiratory mechanism is undoubtedly connected to the fact that the whale has a large number of floating ribs, that is, ribs not attached in front (which in turn means that whales cannot breathe when they come up on land). The lungs are also very elastic, which should help both to increase the efficiency of ventilation and to permit the pulmonary alveoli to change in volume at various pressures, which would reduce the absorption of nitrogen into the blood at great depths. Human divers, as we know, get the bends if they ascend too quickly after a stay at any great depth. In other words, the nitrogen that dissolved in the blood under great pressure is released and forms bubbles in the blood vessels. These bubbles can cause death by cutting off the flow of blood. The whale's ability to avoid this danger depends partly on the construction of its lungs, but primarily on the fact that, as opposed to the diver, the whale holds its breath during its entire dive and thus does not supply the blood with more and more nitrogen.

The problem of providing the body with sufficient oxygen has been solved by means of several mechanisms operating jointly. Blood and muscle tissue have a great capacity for storing oxygen, in addition to which the animal's circulation appears to alter during the dive so that the brain and other organs requiring oxygen receive a greater share than they otherwise would. Moreover, blubber has a high affinity for gases, both oxygen and nitrogen, and can therefore act as a kind of reserve. In this way the blubber probably also helps to solve the problem of the oversupply of nitrogen. It has also been suggested that nitrogen is eliminated by means of the oily foam expelled along with the exhalation of air when the whale blows. The very existence of this foam has been disputed, but for one thing, foam of this kind has been found in the lungs of dead whales, and for another, a whale's blow is clearly visible even in the heat of the tropics. Whalers also report that whale-blow burns on human skin, and that in calm weather whales leave an oily wake.

The sense organs of whales are of course the same as in other mammals, but have evolved to different levels of efficiency. Smell and taste seem to be lacking almost entirely. The sense of touch, on the other hand, is well developed, as we mentioned earlier, while vision does not seem to be of any great consequence. The sperm whale does indeed have very large eyes, about twice as large as those of a cow, but the number of cells in the retina is relatively small. It also appears that whales are color blind. The walls of the eye are thick as a protection against high pressure, and the lens, as in seals (and fish), is almost round, that is to say, designed for underwater vision—which means that in air whales must be extremely nearsighted. In the freshwater dolphins mentioned earlier, who live in muddy rivers, the eye is almost totally degenerate.

Hearing is of course the sense that is most

interesting in the sperm whale. From outside, its ear looks very modest—a hole about one-fifth of an inch in diameter. But inside this hole, the auditory canal is filled with an interesting special formation, a long plug of hardened epidermal tissue with an especially high capacity for carrying sound. This plug rests against the eardrum, from which sound is carried in the usual way via the auditory bones through the middle ear to the inner ear. The whale seems to solve the problem of pressure changes, which is so troublesome for human divers, by means of spongy folds of mucous membrane that project into the middle ear. These fill with blood and swell up when pressure increases, so that counterpressure is maintained. In order to isolate the inner ear from sounds not carried by this plug, the bones encircling the inner ear are not attached to the rest of the cranium and are insulated by foam-filled cavities that bulge out from the Eustachian tube.

As already mentioned, the organs of reproduction have been adapted to streamlining, and even birth and the feeding of the young differ from these same functions in other mammals. The female gestates from 16 to 17 months before giving birth to a calf about ten feet long. Like all whales, it is born tail first, which is undoubtedly the consequence of an aquatic but air-breathing life. For, since the young are so large, birth takes quite some time. There are two mammary glands and nipples, plus a muscle tissue that pumps the milk from its reservoirs so that nursing takes place quickly. Calves are reportedly not weaned until 13 to 14 months, and pregnant females have been found still suckling a calf. Twins can occur but are very rare.

Sperm whales feed primarily on squid. The types they prefer run generally about three feet in length, but small species may also be eaten. The stomach of one sperm whale was found to contain 28,000 specimens of one small squid. In addition, sperm whales catch various kinds of giant squid, about which more later. Even if squid constitute very much the major part of the diet, fish too are often found, from sharks to sardines in size, with deep-sea anglers as an interesting variation. Occasionally we even find fishing gear, where the whale has presumably stolen fish that were hooked fast. We have also found more unusual items such as crabs, lobsters, jellyfish, and sponges, not to mention stones, floats, coconuts, old shoes, and, inevitably, plastic bags.

The sperm whale's diet gives rise to one of its peculiarities, namely ambergris, a dark brown, waxy substance that originates in squid and, with the help of a particular bacterium, can form large lumps in the whale's intestines. These accumulations of ambergris can be amazingly large. The record is held by a piece that was over six feet long and weighed 955 pounds.

Ambergris was once so valuable that it was purchased with its weight in gold, but the price has now declined to about six dollars an ounce.

The high price is due to its remarkable qualities as a binder in the manufacture of perfumes. The production of ambergris should probably be regarded as a kind of disease, since the lumps only occur in occasional animals who have apparently been unable to eliminate the substance in the natural way. However, the individuals bearing ambergris generally look to be quite healthy. Ambergris is not infrequently found where it has drifted ashore, since the lumps float, which indicates that in most cases the whales are able to dispose of them naturally.

For the present, the sperm whale is still indigenous to all the oceans of the world. As with other large whales, ruthless exploitation has reduced it to a fraction of its original numbers, but so far it is still no rarity. Its distribution seems to be limited principally by the supply of squid, and the sperm whale seldom moves north of Iceland in the Atlantic or of the 62nd parallel in the Pacific, while toward the south there seems to be no real boundary. Like many other salt-water organisms, squid seem to be particularly abundant in areas where cold, nutrient-rich water wells up from the bottom, and sperm whales have their favorite haunts in such regions. We therefore find concentrations in the neighborhood of the equator, for example in the well-known area off the coast of Peru and Chile that also provides a living for the guano cormorant, and in corresponding areas off the coasts of South-West Africa and Japan. There is a difference in the distribution of males and females in that the males range further both to the north and to the south. Females, especially when they have young, rarely venture beyond 40° north and 40° south latitude.

CHAPTER FIVE

The Behavior of the Sperm Whale

It is clear that we are comparatively poorly informed about the behavior of most marine animals, and that the reasons behind this lack of knowledge lie not only in the purely practical difficulties of acquiring it, but also in the simple fact that, as land animals, we have a greater aptitude, and probably a greater interest as well, in investigating and penetrating the behavior of those animals that share our own environment. Despite centuries of whaling, we know surprisingly little about the natural behavior of the different species of whales. For obvious reasons, whalers have been most interested in such things as whale migrations, and in whether or not the appearance of whales at certain hunting grounds was regular and fixed. As for other types of behavior, the majority of their observations have involved animals that were frightened and pursued. In other words, we encounter the same difficulties here that apply to any hunted animal—the knowledge of the hunters is indeed extensive but narrowly limited to the information needed for an effective hunt. A solitary Sunday hunter can perhaps afford to settle down somewhere out of pure curiosity and let his prey enact a piece of its natural life before his eyes, but a whaling ship is not likely to lie still for such a purpose.

When the ancestors of the whales returned to a life in the sea, they encountered an environment that in most essential points was utterly different from their home on dry land. Consequently, one rather rewarding manner of approaching cetacean behavior is to try to figure out what demands this new environment made on those first whales and then to look for the behavior and the adaptations that correspond to these demands. We will probably never know the background of the original evolutionary impulse toward aquatic life. The primary cause was undoubtedly the fact that the sea provided a much greater and more uniform supply of nourishment, particularly in regard to plankton organisms and squid, the principal elements in the diets of baleen whales and toothed whales, respectively. That evolution rewarded the animals that "discovered" these sources of nutrition is easy to understand, just as it is hard to explain why none of the sea's older inhabitants ever settled on these specialities and utilized this ecological niche.

The first demand made on a would-be aquatic creature is obviously that its vital functions can adjust to the new environment. Let us take respiration as an example. It may seem to be a serious disadvantage for whales to have to return regularly to the surface to fetch air. But in fact, part of the explanation of the whale's success is to be found in the very fact that it can make use of the infinitely greater oxygen supply to be found in the air. As can easily be calculated, the oxygen consumption of a large creature maintaining a high body temperature (and a high level of activity) in a medium as powerfully chilling as water would be so great that whales would have to drag around enormous gills in order to cope with aquatic respiration. It is undeniably smarter to solve the problem by seeking out the nearby surface once or several times every hour, which was a behavior pattern that involved no disadvantage worth mentioning at the time it evolved. We are probably misled by our own difficulties under water into seeing the necessity of holding one's breath as a serious drawback, and we forget that whales have completely different systems adapted to this purpose, and that respiration can hardly be more of a burden to them than our daily meals are to us.

The transition to aquatic life required no basic changes in other primary bodily functions such as heartbeat, digestion, metabolism, and muscle activity. On the other hand, whales had to adopt an entirely new means of locomotion, though even here the basic design of the mammalian body was good raw material to work from. In fact it is amazing that such small alterations in this basic plan could change the whale from a quadruped to a "fish." The mammalian way of mating has always been well suited to life in the water, and the other aspects of reproduction required no great degree of change. Even breast-feeding, one of the "inventions" that made mammals so successful, is relatively easy to accomplish in water.

The really great and utterly new problems are encountered on a different level, and it is typical of our dry-land lack of imagination that we have not seen these difficulties until quite recently. They can be collected under the headings "orientation" and "communication." A little thought makes it obvious that a creature that moves from land, where its field of vision is endless, down into the sea, where vision is from 60 to 100 feet at best, and in large parts of the ocean altogether nonexistent, must adopt wholly new methods of taking its bearings in this new world. How, for example, is a blue whale that cannot even see its own tail supposed to find a sexual partner in an environment that for all practical purposes precludes both sight and smell? Unless of course the sea is so thick with whales that chance alone is a sufficient tool—or unless it has some other means of orientation and communication. How is the sperm whale supposed to find its prey in the pitch-blackness 3,000 feet below the surface? How are the animals in a herd to stay together when they cannot even see themselves, and how are the young to find their mothers and the other way around? Aquatic life demands new forms of orientation and communication, and it is astounding that our discovery of these new systems should be a sensation in the twentieth century, when even Aristotle was aware of the whale's ability to make sounds. A moment's reflection should have indicated that whales need to keep track of their surroundings, their prey, and others of their own kind, and that sound must be the tool they employ. This omission can only be blamed on our ignorance of the sea and the conditions it sets for life.

As an environment, the sea makes demands that are entirely new in many respects but that nevertheless can be met by means of surprisingly small alterations. It is quite clear that the first and greatest demand made on a land animal when it changes over to aquatic life is that it be able to find sufficient food to support itself. As soon as the physical dimensions of an animal pass certain rather modest limits, it can no longer survive as a stationary organism but must seek out its food. This in turn demands that the animal be able to move freely in its environment, and that it be able to find and capture prey.

The next demand environment makes on behavior is that reproduction be successful. Here of course there is a loophole: the animal can seek out completely different surroundings for reproduction; in this case, it can return to land during the breeding period. Seals and many other aquatic animals actually do so. Oddly enough, the reproductive apparatus of mammals (and other animals) is still in fact an aquatic construction, where the spermatozoa swim in a liquid medium, reminiscent of the time when they swam through the waters of the sea itself. Evolution took great pains to transform these constructions so that they would function in the air instead, whereupon it forced the animal to leave its new aquatic environment and return to the one that had given rise to the transformation of what was originally an aquatic organ!

Leaving the protection of the sea in order to find breeding places on land is a risky step, and it has plagued both seals and sea turtles, especially since man made his entrance on the scene. If aquatic life is to be wholly successful, then the entire reproductive cycle must take place in the sea. But here new problems arise, mostly of a social nature. If individuals no longer gather in a few small breeding areas, then they must be able to find sexual partners in the open sea instead, and they must somehow be able to communicate with them. Also, the young must be given care and protection, which in turn requires some form of social organization, such as the herd. And this puts even greater demands on communication. Thus moving the reproductive process to the sea brought with it a whole series of new problems.

Aquatic life demands new forms of behavior in regard to locomotion, the capture of food, orientation in murky water, a means of finding other members of the same species at great distances and of then communicating with them, plus some form of social life. In the following pages we will look a little more closely at the way these problems have been solved by whales in general, and by the sperm whale in particular.

Swimming and Diving

An undisturbed sperm whale swims at a speed of 2 to 4 knots, which is also the normal speed recorded for herds roaming slowly about within their area of habitation. When they are migrating, this speed may increase by a knot or two, but hardly more. The highest recorded speeds come of course from hunted and frightened animals, who for a period of a few minutes can achieve speeds of 10 to 12 knots, probably more—there are more or less reliable estimates of 20 knots.

When the sperm whale is not hunting but only swimming from place to place, it stays quite close to the surface, either just below the surface itself (if the weather is quite calm), or else deep enough to avoid being disturbed by wave motion. Swimming is of course interrupted by respiration, which occurs at very regular intervals that vary in length according to the size of the animal. An ordinary breathing interval for a large male is actually as much as an hour. Several scientists who have followed male sperm whales on migration report that they have seen breathing occur with clocklike regularity, and that the duration of the dive and the number of blows in each breathing period remained constant for days of uninterrupted swimming at a time. Consequently, we can state that underwater periods from half an hour (females) to one whole hour (males) are entirely normal and correspond to our own normal rate of respiration.

Most whales can swim with a large part of their bodies out of the water. Such behavior is especially common when they are swimming fast, which is explained by the reduction in drag. In addition, the whale's elevated head acts as a balance for the heavily laboring, horizontal flukes and the body, which in this way comes to stand at an angle to the surface, moving through the water somewhat like a plane. However, this method of avoiding some part of the water resistance is only used in emergencies. Water resistance is of course a considerable problem in swimming, and it is very interesting that the whale, like many other aquatic animals, seems to have solved this difficulty by acquiring a body surface that prevents water from forming small drag-producing eddies over the entire contact surface. Instead, the water slips past the body in a smooth and even flow. The secret of this seems to be velvety surface structures that even-out the pressure differences that produce eddies. That this kind of surface makes an enormous difference is apparent to anyone who has ever noticed what an apparently insignificant effort whales, seals, fish, and the like seem to require in order to glide for long distances. Artificial surfaces of this kind have even been tried on boats.

What depth a sperm whale can reach in its dives has been the subject of much discussion. Whalers' reports of depths of 5,000 feet and more, based on the amount of line drawn out by large harpooned males, were long met with distrust and ridicule, since the tremendous pressure should make it impossible for whales to survive at such depths. But indisputable evidence of deep dives was found in 1932 when a 46-foot whale was discovered that had entangled itself in and damaged a cable at a depth of 3,240 feet. In recent years we have also managed to make sound recordings from depths of about 3,300 feet, where whales seemed to frolic in the best of health. There is probably every reason to suppose that the large toothed whales can dive much deeper than these figures indicate, since we know that squid can descend to considerable depths.

The whale spouts after every dive, that is, it blows air and water vapor from its lungs with great force, and then breathes in and out approximately as many times as it has been under water in minutes. Whalers say that a whale of 60 tons stays under water for 60 minutes and then blows 60 times. In such a case, the pause at the surface lasts about 10 minutes. The first blow in a series is the most powerful, a diagonal jet that can be seen at a great distance in cold weather and that reaches a height of from 6 to 50 feet. In children's books and the like, this is often depicted as a shaft of water, but it actually consists of exhaled air, although with a number of interesting additions. These include a great deal of water, especially the water that was in the blowhole, plus what water the stream of air pulls with it from the surface. Another ingredient seems to be the foam we mentioned earlier.

The Quest for Food

We know very little about what methods the sperm whale and other deep-diving toothed whales use for capturing their prey. But since we know that they cannot use their vision, and since other whales that are more easily studied make use of echo-location, the conclusion is obvious. This conclusion is further born out by the fact that in smaller whales the oil-filled cavity in the head, the so-called melon, has been shown to have an acoustical function. Since the sperm whale is the best equipped of all whales in the matter of oil cavities in the head, it seems reasonable to assume that it is also outstanding in the field of echo-location, especially as we know that it can produce sound. We thus arrive at a picture of the sperm whale diving deep into a pitch-black, ice-cold darkness where it uses echoes to search its surroundings for prey.

A number of observations confirm the assumption that the actual capture of the prey is little more than simple swallowing. There is only one eyewitness description of a battle between a sperm whale and a giant squid, but other evidence of the size and belligerence of this prey can be found in the stomachs of captured whales. In 1955, the stomach of quite a small sperm whale, 46 feet, yielded a ten-armed squid that weighed nearly 450 pounds and was 34½ feet long. We have also found arms belonging to probably even larger squid. Last but not least, a large number of sperm whales show deep scars 6 to 10 feet long from the beak and claws of squid, and, in addition, the marks of suckers that in some cases were 4 inches in diameter, considerably larger than those on any squid specimen ever captured. (According to somewhat uncertain reports, the largest squid are supposed to be over 20 feet long, not counting the arms which come to another 30 or 35 feet.)

It appears, consequently, that the sperm whale is a fearless as well as a skillful hunter, a conclusion born out by the fact that a large part of its diet can consist of fish, which must make great demands on the hunter's localization capacity and speed.

By examining the species that occur in sperm whale stomachs, we can get an idea of where in the sea it does its hunting. Two principal regions suggest themselves. First, the waters at those depths inhabited by the lesser squid, and second, the ocean floor, where the larger squid (probably) as well as rays and other bottom fish are found. It has even been suggested that the sperm whale hunts by letting its lower jaw hang down and skim through the sea-bottom ooze, because several specimens have been found with their lower jaws caught in undersea cables. This does not seem unlikely when we consider how difficult it must be to echo-locate flat bottom fish.

Finally, as we mentioned earlier, it is interesting that the sperm whale does not seem to use its teeth in capturing its prey but only in battling with other members of its own species. Any number of well-nourished individuals have been found with seriously deformed lower jaws and worn or injured teeth. We have also found a sperm whale that was completely blind and seemed to have been so for some time, but that clearly had had no difficulty in living a normal life, at least as far as diet was concerned.

Echo-Location

Even if we have not directly demonstrated the use of echo-location by sperm whales, it is nevertheless quite clear that such a mechanism exists. In addition to noting the presence of various specialized constructions in the ear and head, we have succeeded in recording the short, clicking sounds that we know to be echo-signals in other whales. In the sperm whale these sharp clicks have a duration of from 2/1000 to 24/1000 of a second, depending on how many pulses make up the click. Despite the fact that these clicks seem extremely short to us, they actually consist of several shorter constituent parts. The clicks are emitted at rates varying from 1 per second to about 40 per second, considerably slower than the bottlenose dolphin's 800 per second. This difference in frequency undoubtedly depends on the fact that the sperm whale has less need than the smaller whales of being able to investigate its immediate surroundings, that is, the area within 5 feet. The most common rate in a calmly swimming sperm whale is about 2 clicks per second. With this figure as a guide, we can draw several conclusions. When a click has been emitted, the next should not follow so closely as to drown any echo from the first (although it appears that at least the bottlenose dolphin can send and receive at the same time). Thus the animal should adjust its rate of clicking to the distance within which interesting objects can be expected, or perceived. Two clicks per second corresponds to a distance of about 1,000 feet, which sounds like a very reasonable radius within which it would pay to look for prey, especially since squid probably give a weak echo.

Interestingly enough, but not unexpectedly, every sperm whale seems to have its own individual "signature" melody. It is quite natural that each individual should be able to distinguish its own clicks, since these sounds are heard for long distances under water and might otherwise interfere with other animals in the vicinity.

Communication

Practically all we know about the sperm whale's communication is that it exists. Whalers have often observed the way a harpooned or frightened individual can alarm other sperm whales within a range of over 6 miles. They can also make sounds other than clicks, namely creaking, banging, and squeaking sounds, which may lie deep in the bass register. Scientists have recently demonstrated that sounds of this kind are used for communication by the humpback whale. They have succeeded in recording "songs" several minutes in duration, consisting of a low rumbling mixed with rising trumpet blasts. The song is repeated again and again and emitted with sound intensities of 100 to 110 decibels, that is to say, in a class with a pneumatic drill. What is most interesting in this context is that the song was heard at a depth of about 3,300 feet off the east coast of North America, which is just at the depth where we find two sound-reflecting layers close to each other. (These layers exist in all the oceans, at various depths.) Our calculations indicate that a sound of this intensity, emitted within the reflecting layers, could be heard by a human ear at a distance of well over 25,000 miles (which is the circumference of the earth). Even if we make deductions for disturbance, background noise, and the like, it appears that by seeking out this communication layer whales could call to each other over distances as great as the entire breadth of the Pacific Ocean. If the humpback whale can use this system, then there is every reason to believe that the sperm whale can use it as well.

Finally, there are cases of sperm whales having had "conversations" with echo-sounding devices on ships, that is, they have sent synchronized signals in response to the regular beeps of the echo-sounder. They seem to discover their error relatively quickly, but their voices have been registered on the echo tape and can be used, among other things, to determine the depth at which the whale was swimming.

Social Behavior

Ever since whaling began we have known that sperm whales live in small groups organized on the principle of a harem. That is, an old bull is usually followed by twenty to fifty cows and calves. As with land animals that live in this way, we also find herds of half-grown males that roam about by themselves (and, as mentioned above, venture further to the north and south than the harem groups). We also find single males living the lives of hermits. Harem groups can band together into large herds, especially during migrations, but this behavior has grown less and less common since the introduction of whaling. Herman Melville gives an extremely interesting description of such a herd, where the large males swam on the outskirts and formed a kind of breakwater against large waves, creating a calm surface inside where the females could nurse their young undisturbed. Melville's every word is not to be taken literally, perhaps, but he undeniably had firsthand experience of whales. Large herds of this kind gather, or used to gather, during the annual migrations to and from the breeding grounds. Such breeding grounds seem to exist in the vicinity of the Galapagos Islands and of the Azores, but this is a subject about which we know very little. Whalers have long known of these migrations, not least of all because at such times the whales swim too fast to be captured.

Nor do we know very much about other forms of herd behavior, since whalers always split herds up. There are reports of the way whole schools simultaneously dive, blow, and leap, which indicates that some form of social communication must take place. For that matter we have heard click-sequences that seem to have this kind of function. Oddly enough, we also have reports of whalers harpooning one sperm whale after another in a herd without the nearby animals bothering to flee. It even seems to be the rule that harpooning whales does not produce alarm or flight among the other members of the herd, and an annual catch of tens of thousands of sperm whales shows that they are not timid.

As mentioned above, adult males fight ferociously with one another, presumably in connection with the formation of harems. They have been seen to rush at each other with great speed and bite into each other's lower jaws, with quite serious injury as the result. The deep scars and damaged teeth on older males clearly indicates that this behavior is normal and common.

One piece of behavior that is completely unexplained is a curious herd formation described by several observers. The sperm whales gather nose to nose in a large "star" and wafe their flukes. We can speculate as to whether this is some form of communication or protective behavior, but because of our lack of information, the explanation will have to rest there for the moment.

It is interesting to note, finally, that especially in the days of sailing ships, whalers reported that sperm-whale herds almost always fled into the wind once they had been frightened. Presumably, this was because sailing ships could not pursue them at an equal speed, since they were forced to tack. On the other hand, there are also reports of this behavior from the 1940s, when it no longer filled any such function.

Reproductive Behavior

Sperm whales seem to seek out certain special areas when their young are about to be born. It also appears that most mating occurs within these areas, which seems not unlikely since the females often mate while still nursing a previous calf. Copulation takes place belly to belly, either in a horizontal position with the animals lying on their sides, or vertically, where they rise up out of the water. It lasts about ten seconds, according to reports, but has not been observed very often. The calf, which is born tail first, can immediately swim and follow its mother. It nurses lying at her side, and holds the nipple in the corner of its mouth. The female usually turns on her side at such times so that the calf can hold its blowhole above the surface. Growth, as in all whales, is rapid. During the little more than a year that it nurses, the calf grows from 13 to about 23 feet, and reaches sexual maturity at lengths of 30 and 40 feet for females and males respectively. Unfortunately, we do not know how old the whale then is, and guesses range from a couple of years to nine years.

The females do not seem to protect their young with any great intensity, but they will stay close to a harpooned calf. As soon as the calf dies (and communication stops), they flee. Whalers used to take advantage of this fact by shooting the young with a special harpoon, the "grappling iron," constructed so as not to kill the animal but merely inflict the greatest possible amount of pain. With the calf then writhing in agony and crying out for help, they could be sure that the mother would remain nearby so that she too could be killed. This scheme definitely gives us a certain insight into the mentality of whalers and their attitude toward their prey. Unfortunately, there is no reason to suppose that these attitudes have changed, even if the hunting of calves is forbidden at the moment for economic reasons.

Thus even if the female does not go to the

attack, she and other members of the herd react strongly to the calf's distress signals. The well-known French skindiver, Jacques Cousteau, was following a family of sperm whales when, in his eagerness to get good photographs, he went so close to the animals that the propeller of his boat ripped open the side of a calf. Within a short time, twenty-seven other sperm whales showed up out of nowhere and gathered around the injured animal. Similar incidents have often been observed. Males do not appear to be particularly interested in the young, but will protect both them and the females if the harem herd is threatened. In some cases, females have also been seen to stand by harpooned males, but as a rule they leave them to their fate.

Behavior Toward Man

As can be seen from the above, the sperm whale shows no great fear of man. This phenomenon is not unique, since there are animals that actually seem to lack the capacity for fear, namely, such species as are not threatened by any predators or other natural enemies and therefore have never been exposed to any selection pressure for the development of flight behavior. Of course this is not entirely true of sperm whales, who both can and do flee once they have been harpooned, but all the same it is astounding that after centuries of merciless butchery, these animals still let human beings come close enough to harpoon them. There can hardly be any doubt but what a sperm whale, with its great speed and underwater endurance, could easily escape from a whaling ship if it really "wanted" to. This is unquestionably true of the days when the hunt was conducted from rowboats. The probable explanation of the fact that this and other species permit themselves to be slaughtered is quite simply that they have never been equipped with any flight mechanism released by enemies on the ocean surface. They have no such enemies in nature, and have certainly never had them. The animal kingdom can provide many parallel cases showing an inability to adapt flight behavior to human hunting methods, which appeared so suddenly on the evolutionary scale.

A woodcut from Beale's *Natural History of the Sperm Whale*, 1839, showing sperm whales being hunted with hand harpoons.

Perhaps it is also amazing that whales so seldom attack whalers. The sperm whale is the species that made the best name for itself in offering resistance, particularly when whaling was carried out from open boats, which wounded whales regularly bit in two or crushed with their flukes. Whalers used to divert the attacks by throwing out a cask, which the whale then attacked instead of the boat. Apparently it was not uncommon for companions of the harpooned animal to gather around it and attempt to bite off the harpoon or the harpoon line, which they often managed to do. Curiously enough, the harpooned sperm whale seems incapable of perceiving that by taking the line in its own mouth it could quite easily escape the whaleboat. This is also interesting in light of the fact that whales, who after all lack hands and similar implements, are unable to rid their own bodies of unpleasant objects. Instead, individuals seem to help "groom" one another, and the incident just mentioned indicates that sperm whales too have some such social behavior pattern. The conclusion to be drawn from the various observations of cetacean behavior toward man seems to be that, in general, whales lack the ability to comprehend the danger human beings represent. Their natural behavioral equipment is not constructed to cope with a threat from the surface, that is, they will never learn that danger threatens from small (or large) ships and that they could avoid this danger by diving and swimming away under water.

Is the Sperm Whale Intelligent?

It is appropriate to conclude this chapter with some reflections on the sperm whale's intellectual capacity, from a zoological point of view. We cannot escape the fact that the sperm whale is equipped with an excellent brain, which must have developed under the pressure of natural selection and which, like all other organs, must fill some need and be adapted to it. Moreover, we know that the bottlenose dolphin and other small whales display a capacity for learning tricks and solving problems that is not inferior to that of any other animal. But the zoologist must also ask himself why the sperm whale needs a large brain. It seems to live a life that corresponds in most respects to that of a large grazing animal on land, the cow for example, and there is no reason that its quest for food should require a great capacity for solving problems and adjusting to new situations. It is hard to see in what way the life of the sperm whale demands any appreciable intelligence in the individual members of the species. On the other hand, we know fairly little about this life, so the argument is not perhaps so terribly convincing. The "minus column" must also definitely include the sperm whale's curious inability to simply stay out of the way of the predator man, particularly during earlier whaling eras when flight would have been a simple matter. Nor does its social life seem to be especially demanding intellectually.

The harem pattern, for instance, is well known in many land animals such as deer, who do not for that reason appear to be conspicuously intelligent.

Why then is the brain so large? One explanation may be the one presented in another part of this book, namely, the possibility that orientation and communication by means of sound require a much larger "data bank" than in other animals. The large brain would thus be needed for echo-sounding and communication, for it is entirely possible that it is very complex and requires a large cerebral capacity. Of course, for the time being it is hard to say why the world of the whale should offer so many more subjects of conversation than the corresponding environment of land animals, but of that we know very little.

We are forced to state that what we know of the sperm whale's behavior gives us no cause to consider it more intelligent than, say, an elephant. But then the question immediately arises: What do we mean by intelligence? A little thought will tell us that what we take such pride in—our capacity to adapt to new situations and to use previous experience to solve new problems—is an ability that harsh selection called forth in a naked ape turned predator. And this ability is suited to the life of such an ape, even if it has also lent itself to the invention of whaling ships, hydrogen bombs, and concentration camps. With a little imagination one can actually conceive of a different type of intellectual activity than ours. And it need not necessarily be the case that our form of intelligence is the most appropriate form for whales as well. For the time being, it looks as if whales are very much our inferiors, when measured by our standards. But maybe it is futile to make comparisons at all. Whales have a large brain capacity, and physiologically their brains function in the same way ours and those of all other animals do. But this does not mean that the intellectual activity must be the same.

CHAPTER SIX

The History of the Sperm Whale

The conventional approach to the whale's history is to describe man's history in relation to the whale. This kind of a history begins with the first pictures human beings made of whales; it describes how we eventually learned to kill whales, how our hunting methods improved over the years, how we learned new ways to make use of their dead bodies. It ends by telling how few whales there are left to kill, usually with a sigh that it would be dreadful if these splendid beasts were exterminated.

An unconventional way of writing the history of the whale is to try to ascertain whether the whale can be said to have a history in relation to man. Unfortunately this kind of whale history turns out to be a little thin, because there are so few facts to work with. Here we shall make use of both methods: quickly describe man's history in relation to the whale, and then speculate a little on the whale's history in relation to man.

Human History in Relation to the Whale

Undoubtedly the most charming and still one of the most complete accounts of human history in relation to the whale is to be found in Melville's *Moby Dick*, from which much of the following material is taken. The whale's adaptation to an entirely aquatic life is so complete that it is quite natural that for a long time men viewed the animal as a fish. We know that Aristotle understood that whales were not fish but mammals. This knowledge was forgotten. The Romans thought of whales as fish, and this continued to be the general view until 1671, when John Ray suggested that whales were mammals. It was Carl von Linné who clinched the argument once and for all when in 1766, in his pioneering *Systema Naturae*, he declared, "I hereby separate the whales from the fish." Among his reasons, he mentions that whales are warm-blooded, have a heart with two chambers, have movable eyelids, and nurse their young.

The classification of the families of whales caused great difficulties for a long time. Scoresby, a well-known naturalist, wrote in 1820, "No branch of zoology is so much involved as that which is entitled Cetology." And in 1839, Beale, another famous naturalist and ship's doctor, stated, "It is not my intention, were it in my power, to enter into the inquiry as to the true method of dividing the cetacea into groups and families." The famous zoologists and anatomists Cuvier, John Hunter, and Lesson agreed that it was impossible to complete the study of whales "in the unfathomable waters." According to Melville, they refer to an "impenetrable veil covering our knowledge of the cetacea," and call cetology "a field strewn with thorns," where a variety of "incomplete indications but serve to torture us naturalists."

We were well into the twentieth century before cetology left the realm of wild speculation and approached reality. Even today our kowledge of whales shows great gaps on very elementary subjects. As we mentioned earlier, there are species of whales that have only been seen in the form of badly preserved stranded individuals, and we have a very poor idea of how many specimens of these rarely seen species there may be. They may exist in much larger numbers than we imagine, but are able to stay away from man and his whaling ships.

Man and the Sperm Whale

Many of the whales described in the literature of the Middle Ages and even later as hideous, gigantic monsters, savagely attacking seafarers and splintering large ships in their enormous jaws, must have been sperm whales.

Olaus Magnus's famous *Carta Marina*, printed in 1539, depicts a number of these appalling beasts, crushing ships and unfortunate seamen. One of these sea monsters is almost certainly a sperm whale, if rather imaginatively drawn. Because of its peculiar appearance, the sperm whale has always been easy to distinguish from other whales. A little below the sperm whale Olaus Magnus has drawn a lump floating in the sea and furnished it with the label "Ambra Spermaceti."

Of course, it was not just a thirst for knowledge that lay behind the intense interest in these strange animals, but rather foremost and for centuries, a crass economic thirst for wealth. Here was an apparently boundless supply of meat to be taken from the sea. And not only meat but also valuable oil and a whole line of other products. Whaling developed quickly into a worldwide industry, a ruthless butchery that threatened to eradicate several species.

The flesh of the sperm whale was considered bad-tasting and inedible, which meant that for a long time the animal was left in peace, despite the fact that it usually swims quite slowly and is relatively easy to catch. It was not until the eighteenth century, when there was a great increase in the demand for lamp and candle oil, that the sperm whale became an attractive quarry for whalers on account of the great quantities of oil it contained.

At the same time, the myths and legends began to grow around this dread monster of the deep. It was the time of Moby Dick, the great white sperm whale that inspired Herman Melville to write his famous novel. A few lines from *Moby Dick* deserve to be repeated here.

Melville writes that on the subject of the Greenland right whale, Captain Scoresby is the best authority. But, he goes on,

> Scoresby knew nothing and says nothing of the great Sperm Whale, compared with which the Greenland Whale is almost unworthy mentioning. And here be it said, that the Greenland Whale is an usurper upon the throne of the seas. He is not even by any means the largest of the whales. . . . Reference to nearly all the leviathanic allusions in the great poets of past days, will satisfy you that the Greenland Whale, without one rival, was to them the monarch of the seas. But the time has at last come for a new proclamation. . . . Hear ye! good people all,—the Greenland Whale is deposed, —the great Sperm Whale now reigneth!

The pursuit of the sperm whale fell off during the second half of the nineteenth century, and the species had a chance to recover from the massive slaughter that had gone before. The hunt was resumed in the twentieth century, but it never reached the same proportions as earlier. Nevertheless, the catch is still considerable. In the 1957–1958 season, a total of 18,853 sperm whales were taken. An agreement has been reached to limit the catch to animals over 38 feet, which looks as if it will spare a sufficient number of females to maintain the population at its present level.

The History of the Sperm Whale in Relation to Man

It is hard to imagine that the sperm whale has anything we could call a history in relation to man, unless it is capable of rational thought on a social level. In order to speculate about such a history, we must simply assume that the sperm whale has this capability. Observe that we do not maintain that the sperm whale *does* have this ability! The reader should look at this section as an intellectual experiment based on this wholly unproven assumption, and as a continuation of the discussion of whether or not the sperm whale is intelligent.

In what for man were prehistoric times, the sperm whale was the absolute ruler of the sea, to all appearances a mild-mannered ruler who specialized in catching giant squid in the depths of the ocean. Obviously they had nothing to fear from the larger baleen whales. Nor did healthy animals have anything to fear from killer whales or sharks. And Stone Age man couldn't do them any injury worth mentioning.

Human beings did not become any great danger until the 1700s when whaling began to be conducted from sailing ships, from which small boats with harpoons were set out. Up until that time, the only products of our culture the sperm whale came in contact with were different kinds of boats. As experts on squid and other sea animals, they could hardly be interested in human beings as food. Nor is there anything to indicate that the sperm whale took any interest in killing human beings prior to the eighteenth century. Of course, it is possible that occasional sperm whale psychopaths may have appeared and made trouble for mankind even before the days of whaling. A historian by the name of Procopius, who lived in Constantinople in the sixth century after Christ, speaks of an enormous sea monster that sank boats in the Mediterranean at irregular intervals for roughly fifty years. Occasional sperm whales have been observed in the Mediterranean, and this sea monster may have been one of them.

When the great whale slaughter began, we can assume that whales wanted to defend themselves against man. They could either accept the challenge and initiate a common defense in the form of attack, or else they could choose to stay out of the way as much as possible. All the available information indicates that whales continued to leave human beings in peace even after we had begun to do them so much harm. Of course this lack of reaction may have been unconscious, but then again it *may* have been conscious.

Individual misanthropic sperm whales have been reported ever since the beginning of the eighteenth century. Moby Dick was not the only monster that "bellowed in the ocean's wild sway," to borrow Erik Axel Karlfeldt's description of the whale that swallowed the prophet Jonah. There were others who were equally feared: Timor Tim, New Zealand Jack, and Newfoundland Tom, to name a few, all equally legendary for the devastation they visited on seamen, boats, and even large ships. They could really be very dangerous when they were fighting for their lives. It was not all that rare for them to ram boats with their enormous flukes or to splinter them with their huge heads. And the seaman who happened to wind up in the jaws of a whale was quickly crushed by its sharp teeth. Slijper tells of a Dutch ship of 714 tons that was rammed by a sperm whale with the result that the propeller was destroyed and the ship had to be towed to Melbourne for repairs.

By and large, however, the sperm whale has remained remarkably peaceful toward man even in recent times.

The industrial revolution in the human world involved changes in the sperm whale's world as well. It was periodically forced into contact with our reckless discharge of waste and poisons into the sea. Tremendous quantities of mustard gas, nerve gas, oil, and other rubbish cannot have passed the whale unnoticed. And the sperm whale in particular has suffered from our progress in shipbuilding. They often lie resting at the surface, and many have been injured or killed by high-speed ships.

The First and Second World Wars must also

have resounded in the world of the whale. If the sperm whale can think, then it must have been shaken more than just physically by our improved techniques in bombing.

Can the Sperm Whale Be Convicted of Stupidity?

A natural question under the heading of the sperm whale's history in relation to man is whether or not the whale can be accused of stupidity. We have established the fact that so far the sperm whale has maintained a calm and passive attitude toward man. If it has been able to make a choice in this matter, then of course this choice has been a wise one. If the sperm whale were to counterattack and destroy small human ships that have so far posed no danger to the whale, we know what would happen. The human race is resolute and would certainly not put up with this form of insecurity on the seas. The fact that whales remain peaceful toward man cannot be used as an argument either for or against their intelligence.

But what about the sperm whale's apparently characteristic inability to perceive the danger that mankind represents? Isn't this a sign of a failure to adapt, in other words, of stupidity, as we suggested in the preceding chapter? It is possible that this is the case, but it does not have to be the case.

In the first place, it may be that during a surface hunt the animals simply do not receive the information they need in order to flee in what we would consider a rational way. A whale surfacing after a dive must replenish its supply of oxygen during ten critical minutes before it can begin another dive of normal duration. While surfaced, the whale's ability to send and receive sound is entirely or partially nullified. As we already know, its vision is not well developed, on top of which it is nearsighted in air and has a very limited field of vision. Blind and deaf, it becomes an easily captured quarry on the surface. It is in deep water that the whale comes into its own. It would be interesting to see how a submarine would make out as a whaling ship.

Another aspect of the sperm whale's amazing passivity during the hunt has to do with our own anthropocentric point of view. Even if the whale does receive enough information to allow it to act rationally by our standards and yet does not do so, that does not necessarily mean that it is stupid. The fact that we cannot explain its actions is insufficient evidence, as we pointed out in the preceding chapter. An intelligent being from another planet would no doubt seriously question the rationality of man. Naturally by *our* standards we have "rational" explanations for all the idiocies we have committed and are now committing, but a creature with a different emotional background might have trouble seeing what was rational about, say, religious and political ritual murder, or about the fact that in olden times women used the baleen from right whales to obstruct their breathing and reduce their freedom of movement.

Has the Sperm Whale Demonstrated Its Intelligence?

We cannot produce any definite proof that the sperm whale is intelligent in our meaning of the word. Actually our knowledge of the whale is very incomplete. All we can do is repeat an account of curious whale behavior that a generous interpretation might label intelligent—if in fact it actually occurred.

In *The Cruise of the Cachalot*, first published in 1897, Frank T. Bullen relates an incident from the latter part of the nineteenth century. The observation was made from a whaling ship whose hold and deck were already so filled with the spoils of a successful hunt that there was no room for more. A school of sperm whales began "disporting all around the ship, apparently conscious of our helplessness to interfere with them," Bullen writes. He goes on to tell how the crew was allowed to witness an uncommon spectacle.

> The whole school surrounded the ship, and performed some of the strangest evolutions imaginable. As if instigated by one common impulse, they all elevated their massive heads above the surface of the sea, and remained for some time in that position, solemnly bobbing up and down amid the glittering wavelets like movable boulders of black rock. Then, all sud-

denly reversed themselves, and, elevating their broad flukes in the air, commenced to beat them slowly and rhythmically upon the water, like so many machines. Being almost a perfect calm, every movement of the great mammals could be plainly seen; some of them even passed so near to us that we could see how the lower jaw hung down, while the animal was swimming in a normal position.

This spectacle, which must have been a fantastic sight, lasted for over an hour until the animals suddenly disappeared, again as if on command.

Bullen's account is from 1897, at which point human beings had been ruthlessly slaughtering sperm whales for nearly two hundred years. Why did the whales put on this curious performance for a whaling ship overflowing with its catch? The whales cannot possibly have been unaware of the ship's presence. It is possible that there is some explanation that is natural in the conventional sense and that does not assume that the whales are thinking creatures. We cannot find it. But we can supply an explanation based on the supposition that the sperm whale is capable of rational thought on a social level. It is possible that the whales somehow understood that this time they would not be attacked, or at least that the risk was small. They took the chance of presenting this odd display, whose purpose was to draw human attention to the fact that sperm whales are not simple, ordinary, stupid fish but highly unusual creatures, who do indeed lack hands with which to make harpoons, but who have other qualities, among them the ability to cooperate in putting on this otherwise utterly gratuitous exhibition.

The future will show whether this explanation is too imaginative. (Or was it Bullen who imagined things?) One also wonders if the curious behavior of the whales in a herd when they make a "star," which has been reported several times, cannot be explained in the same way.

PART III

Man and the Tree of Knowledge

CHAPTER SEVEN

A Complex Relationship

Man's Dominance Is of Recent Date

Only a few hundred years ago, the sperm whale was the dominant or at least an unconquered species in the sea. At the same time, man was the dominant species on land. Another 10,000 or 100,000 years earlier, before man learned to lure elephants and lions into traps, these animals too were unconquered and dominant in their own domains. By virtue of his culture, but not necessarily only by virtue of his intelligence, man has come to prevail virtually everywhere on earth, in recent years even in the sea. This is a unique situation—that one species should have achieved such absolute supremacy. Mankind's problem is to adapt to this situation.

The facts we have tried to emphasize in this book affect people's thoughts and actions in many different ways. Some people grow humble when presented with the information that our dominance is of so recent a date. Others are unimpressed by the time factor, content with the knowledge that we are now worst in every way.

Man Is an Animal and Ought to Be Studied as Such

It was not until the end of the nineteenth century that we began to demand "real" objectivity in our study of animals. One of the people who were most important in this connection was C. Lloyd Morgan (1859–1936). He tried to make a distinction between the methods of natural science and the older, careless, anecdotal approach. He tried to do away with the anthropomorphic fallacy. In 1894 he wrote, "In no case may we interpret an action as the outcome of the exercise of a higher psychical faculty if it can be interpreted as the outcome of the exercise of one that stands lower in the psychological scale." We must now, however reluctantly, admit that man is an animal. And that being the case, it is only reasonable that we apply the critical methods recommended by Morgan to the study of man as well.

Man Is Naturally Aggressive Within His Species

Almost all animals are aggressive toward other animals of the same species and toward animals of closely related species. This aggression has a positive value for the survival of the species as a whole. Konrad Lorenz analyzes this phenomenon in his book *On Aggression*. Human beings display undeniably aggressive behavior in the sense that they often get angry. Some sociologists and psychologists (fewer and fewer, fortunately) maintain that man is not naturally aggressive but has learned to be so, that aggression is entirely culturally determined. For a biologist, the probability is overwhelming that man is naturally aggressive toward other members of his own species, that we are dealing with a "both-and" rather than an "either-or" phenomenon. We know that function is the leading factor in phylogenetic development. Since the similarities are so great between animal and human aggressive behavior, it seems highly unlikely that the basis of *our* aggression should be so different. The burden of proof lies on those who assert that man's aggression is entirely learned.

Still, it is quite clear that we learn the manner in which we show our aggression, and that we can learn to repress or express our inherent aggressiveness within certain limits. From this point of view, it is nonsense to speak of "the instinct for war," since war as it has been fought in historic times is so much a product of the cultures we have built up—that is, a product of what we have learned. In fact it is only the most extreme behaviorists who accuse ethologists of talking about the "instinct for war"—no modern ethologists use that term.

Without committing ourselves on the question of how much of our aggressiveness is inborn and how much is culturally determined, we can certainly vouch for the fact that man has shown great efficiency in the annihilation of closely related species. The most humanoid apes and all but one of the human species are extinct. *Homo sapiens* has won. The current genealogical trees of the evolution of man and the modern apes show how effective we have been in eliminating the competition. As apes, we are indeed unique.

The Discontinuity Hypothesis

The question of how unique we human beings really are, compared with the animals from whom we are most nearly descended, leads us to the so-called discontinuity hypothesis. According to this hypothesis, human mental activity is an outstanding example of a so-called "emergent" phenomenon—the appearance in evolution of phenomena that are qualitatively different from previous stages of development, in other words, qualitative evolutionary leaps. For me, belief in or rejection of the discontinuity principle is not important for a total biological perspective on mankind. We know that our brain doubled in size in the course of about two million years—from 600–700 cubic centimeters in *Australopithecus* to 1,400 cubic centimeters in modern man. It is possible that qualitatively new functions came into existence during this period. Language is considered to be one such function. However this does not prevent the great majority of functions from showing continuity. Our aggressive instinct is in all probability a continuing function of this kind. I believe our capacity for devotion can also be traced a long way back, via *Australopithecus*.

Man—a Defender of Territory

Most higher animal species have an innate impulse to defend some definite area as their own, either permanently or for a certain season. This area, or territory, can be private but can also be held in common by several individuals, by a family, or by several families composing a "biological nation." Territorial occupation has the very greatest importance for the continued existence of these different species. It is perfectly obvious to anyone who will stop and think for a moment that man too is a defender of territory. On the other hand, it is very unclear how much of our territorial behavior is built into our genes and how much of it is learned. Those who maintain that such behavior is entirely learned must bear the burden of proof. No biologist insists that man's territorial behavior is completely genetically determined.

It is true of all territorial struggles that the territory-defending individual or group gains moral courage and strength toward the center of its territory and is therefore practically invincible there. Conversely, the territorial defender declines in courage and strength as he approaches his frontiers. Defeat awaits him should he venture outside the limits of his territory, where his battle morale is low and where he would rather flee than fight. If the territories of a given species are made large enough by artificial means, hostilities cease. If we could implement birth control over the entire earth and thereby make our social territories so large that everyone could eat his fill, so large that aggressiveness at the borders was reduced to a minimum, then perhaps war would not break out under circumstances that otherwise seem to trigger it.

Vacuum Activities

But the people want not only bread—they want circuses as well. One might almost believe that Caesar had read Erik Fabricius's textbook on ethology: "If an animal does not come into contact with the sign stimulus for a certain instinctive action over a long period of time, the threshold of response in the animal's releasing mechanism can in some cases rise so high that the threshold value is surpassed and the action breaks out in an explosive manner without demonstrable external stimulus." The classic example of this kind of vacuum activity is Lorenz's description of the behavior of a tame starling that he kept in his house. This bird often flew up to the arm of a chair where it sat and reconnoitered, like a wild starling looking for insects. Then it would fix on some point in the air near the ceiling, fly up and make a snatching movement, return to the chair and strike its bill against the arm as if killing an insect, and then make swallowing movements. When the captive starling was not given the opportunity to catch insects, the threshold of response in its release mechanism for these insect-catching actions finally rose so high that this whole complex chain of behavior broke out by itself, despite the fact that there were absolutely no flying insects in the room.

It doesn't seem at all farfetched to construe much of the violence in our urban world as partial vacuum activity. Man's need for action has to find a release. Our Haymarket riots had complex causes, of course, but I don't believe we would be too far off if we supposed that it was partly a question of vacuum activity in people who, for various reasons, were idle. An earlier generation had pitched battles between collections of schoolboys. The disappearance of this phenomenon has been credited to organized sports. The village fray is ritualized in Little League baseball and ice hockey. By the same token, international championships can be regarded as ritualized war.

International Soccer Matches as Ritual War

An international soccer match is quite an excellent example of ritual war. (I am paraphrasing Robert Ardrey.) First the national flags, and kings or presidents or ministers who greet the players, and then the national anthems. The battle surges back and forth across midfield, the international boundary. The attackers put pressure on the defenders, and the object of the struggle is to send the ball with a fatal blow directly into the enemy heart, the goal. When the contest approaches one goal, the team defending it gains several advantages. The opposing team is forced to leave several players at home to guard its own goal against sudden attack, the goalie is permitted to take the ball in his hands, and the offsides rule hinders the enemy advance. The situation is the same in war—the defender has several trump cards in his hand. At least that's the way war *used* to be.

Violent emotions are released in the course of a soccer match, among both the players and the public. The situation is especially inflammable if the home team loses. Trade agreements have been broken after disastrous matches, and certain players have come to be looked upon as national enemies. The accounts in the newspapers are full of the terminology of war.

Aggression Displacement—an Unexplored Field

The questions surrounding aggression displacement are obvious. Can it be defined and measured in any sort of dependable manner? How much of the stimulus of ritual war can we tolerate? At what dosage does ritual war cease to be aggression-displacing and begin to provoke destructive aggression? In southern countries, fights at soccer matches occur quite often. In May 1964, 319 people were killed at a soccer match in Lima, Peru, and in the fall of 1967, 42 persons were killed at a match in Turkey. But anything is better than war, and we must pursue every possible means to prevent its outbreak. Perhaps a tremendous force of policemen or soldiers at sporting events would strengthen the feeling of ritual war and increase the aggression displacement.

It is very possible that there are other and better methods of displacing aggression. Could we, for example, train ourselves to use short-circuit behavior to displace aggression? When a gorilla doesn't know whether to fight or flee, it starts scratching itself or pretending to eat. We do roughly the same sort of thing. Will it help to sit at the border and smoke a peace pipe, or to scratch our chins, or to chew gum? I am not qualified to discuss the difficult problems of aggression displacement., However these problems tempt me to ask one more question: Have scientists and politicians given this subject the priority it deserves?

Aggression Displacement or Polarization?

Certain political circles take no interest in displacing aggression and are quite indifferent to what we call manipulation (unless it is outward bound from their own circle). They consciously strive for polarization, an increase in the culturally determined antagonisms between different groups of people. Their rationale is that a polarization will hasten the economic and political changes that must come about before mankind's serious problems can be solved. I heartily agree that vast political reforms are needed all over the world, and that some of these reforms will probably have to be carried out by force, since force is being used to prevent them. But polarization is risky in these days of the arms race. As we work for reform, we must constantly ask ourselves one anxious question: What have we been able to discover in the course of our brief history about man's constitutional limitations, both as an individual and in groups?

In the Russia of the 1950s, Lysenko dominated the science of genetics with his thesis that acquired characteristics could be inherited. This doctrine suited Communist beliefs, and all arguments against this scientific nonsense were suppressed. Talk about "human nature," about our constitutional limitations, is not popular among the people who want to build the ideal dream society in the spirit of Marx and Lenin. They are not satisfid with the fact that man is enormously willing to learn, and that at every moment of his life he is a product of inheritance and environment.

The odd thing is that it is precisely the Marxist-Leninists who, on one point, have such tremendous respect for man's constitutional limitations. And this point is our inability to avoid violence. They often cite examples to show that the powerful never relinquish power and privilege until blood has been spilled, but they quietly disregard those cases in history where conflicts were solved without bloodshed. Perhaps all of these bloodless settlements involved the threat of violence, which is a kind of violence. So what? For the parties involved, there is nevertheless an increasingly important difference between the threat of violence and violence itself. I suspect that democracy, even in the form in which it is practiced in Sweden today, is a valuable political instrument for altering society without violence.

More Things Uniting Men Than Dividing Them

The earth at present holds a number of more or less pure human races. The average differences between the races are considerably less than the individual differences within the same race. In other words, there is more uniting us than dividing us. This knowledge makes it easier for me personally to look at all human beings as *Homo sapiens* rather than as members of a certain race or a certain social circle. But it may be biologically natural for us to direct our aggression against what is almost alike but not really alike. After all, it is the "almost alike" with whom we compete for living space. If this is the case, then we must get a grip on ourselves and go against our nature in this regard. Fortunately, our nature also includes thought and a large brain. Our developments in weaponry mean that we can no longer afford the martial extravagance in which we used to be able to indulge without endangering our species.

Now the reader may think that I am beating the drums of doom a little hard. But we mustn't forget that the fantastic developments in communications have made the world shrink. The perspectives are different for the have and the have-not nations. The world of the have nations has become very small, and the world of the have-not nations has an end, whereas it used to be endless. It is no longer possible to get away from the fact that at least half the people in the world are starving, while the rest are dieting. They know and we know. The ease with which biological weapons can be used may mean that the people in so-called responsible positions no longer dare let the starving live. Large-scale organized sabotage against a great power might be misunderstood and trigger an atomic war.

The Rich Peoples Have a Choice

The world's wealthy peoples face a choice: either we will kill the poor as quickly as possible, or else we will decide to look upon the starving as fellow human beings, and act accordingly. Of course it is perfectly obvious that we must choose the second alternative and regard other people as our brothers. That is, it *sounds* obvious, but history suggests that it is far from obvious that we will actually take this view. In fact, it is unlikely. It is so easy for us to find grand and noble reasons for killing our fellow men—as always. This capacity for commitment is written into our genes.

An Unwelcome Doomsday Sermon

Many people are unwilling to see that the situation is as serious as I picture it. And those who cannot see it are not willing to take adequate steps to avert catastrophe. For that reason I would like to dwell on this unwelcome doomsday vision and try to convince the skeptical. In doing so I will make extensive use of a paper by John Platt, an American biophysicist.

The world today faces a series of impending crises. All the threats we see in every direction have their roots in rapid change. The rate of change has increased enormously in the last hundred years. We all know this; it is repeated *ad nauseam*. And yet perhaps we do not realize *how* rapid and radical the changes of the last hundred years have been, in comparison to corresponding changes in earlier phases of history.

During the last hundred years, we in the developed countries have increased

1) the swiftness of communication 10 million times,
2) the speed of travel 100 times,
3) the rate of data processing 1 million times,
4) the means of exploiting energy resources 1,000 times,
5) the power of our weapons 1 million times, and,
6) the possibility of controlling disease perhaps 100 times.

Obviously, such changes present us with tremendous problems of adaptation. A good number of these changes have come about during the last thirty years. This means that people under thirty, who grew up along with a number of these changes, have acquired expectations and concerns unlike those of an older generation that grew up in a different, more slow-moving world.

Another important aspect of these technical changes is that in certain cases they seem to be approaching their natural limits. Perhaps the S-curve is starting to level off. Maybe communications cannot get much faster than what TV makes possible right now. And from the human point of view, weapons can hardly get much worse than they are today—we already have an enormous "overkill" capacity. This means that if we could use the next ten years to learn to master the forces we have set loose, without blowing each other up, we might be able to achieve new, effective social structures that would last for generations. Perhaps we could move into a new world of abundance and variety, made possible by technical progress.

The human race can be likened to a group of miners leaving their mine by way of a tunnel. Suddenly falling rock begins thundering down behind them. At the same instant, they catch sight of the light at the mouth of the tunnel. It's worth making an attempt to reach the light and the air by starting to run.

Various types of crises can reach the exploding point in the next ten years. Nuclear war can break out, worldwide famine can set in, crises can arise in the functioning of democracy: administrative crises, racial crises, and so forth. It will be worth our while to take a slightly closer look at some of these impending crises, so that we may better understand how little time we have to avoid them.

We can begin with the risk of nuclear war. As long as we cannot create stabilizing social structures that will preserve peace in the world, we will go on living under the constant threat of nuclear weapons. Confrontations between the nuclear powers have occurred every other year or so—Korea, Laos, Berlin, Suez, Cuba, Vietnam, and so forth. MacArthur wanted to use nuclear weapons in Korea. During the Cuba crisis,

Kennedy estimated the risk of a so-called nuclear exchange (what a seductive expression!) at 25 percent.

Nuclear war can be triggered by accident, either a technical accident such as a radar failure, or a political accident, where some politician (probably a dictator) is in trouble and sees war as his only personal salvation. But worst of all is the fact that a nuclear war can be triggered precisely as planned—from the border incident, the strategic maneuverings, threats and counterthreats, all the way through to the projected counterattack, which could burn off the entire surface of the earth in a three-hour duel.

What is the likelihood that this will happen during the next twenty years? It is impossible to say, of course. But it is clear that another five or ten confrontations in nuclear roulette will reduce our chances of surviving beyond 1980 or 1990.

The next threat on our list is worldwide famine. Many agricultural experts believe that such a famine will begin during the 1970s, with hundreds of millions of victims in thickly populated areas such as India and China. Other experts disagree and claim that the new types of grain developed over the last few years will permit food production to keep pace with population increase. This will postpone the problem for a time. The pessimists, on the other hand, claim that it takes too long to teach farmers to change their ways and to get consumers to use the new grains, so that food production during the next ten years will fall behind anyway.

But *if* famine arises, it can easily become catastrophic. In addition to the suffering it will cause its victims, it will increase international instability, with riots to get food, the use of troops to suppress the riots, governmental crises, international interventions, and so forth. By comparison, Vietnam might look like a popgun—not to the Vietnamese, of course, but to, say, the Swedes.

The seventies will probably bring administrative crises of various kinds. The administrations of universities, labor unions, cities, and countries won't have time to deal with all the problems. People are protesting already, all over the world. They refuse to accept solutions supplied by a central elite. At first glance, these protest movements may not seem so dangerous next to the greater threats of famine and atomic war. But if we think about the way small instabilities can imperil the stability of the system as a whole, then it becomes clear how important it is for all administration to function. We need control at all levels, from the top down, and from the bottom up.

Most administrations are capable of handling one crisis at a time in a passable way. But when the crises come one on top of the other, things can get bad. Take New York City in 1968: teachers' strike, police strike, sanitation workers' strike, and a dock strike, all within a few days. Under such circumstances it's easy to fall a couple of crises behind. And each individual problem can get worse as a result of the fact that the people who are supposed to deal with it cannot think clearly, do not have time. What would the Cuba crisis have become if it had occurred at the same time as the great power failure on the east coast of North America?

There is no doubt but what administrative crises lessen our chances of surviving beyond 1980. And we know that so far these various problems are still growing worse. The probability is great that ten years from now an estimation of the risks of the coming decade will be even more pessimistic than it is today. One need only think of pollution and the population explosion.

The value of this kind of scare-prediction is not only its capacity to frighten us, but also the fact that it gives us a rough idea of how much time we have. Time is short, but maybe not too short if we increase the pace of our efforts. The future depends on what we do. The quality of our decisions and actions now can make it worse, or better.

Man as Ostrich

They say that when threatened, the ostrich stuffs its head in the sand so it cannot see the danger. Apparently this isn't true. I have searched ethological literature in vain for the phylogeny of this reaction. It appears, however, that man is equipped with a very nearly unbelievable capacity for ignoring reality whenever it seems to be unpleasant. One of the best examples I know of this mechanism is from the Second World War.

Treblinka, which is less well known than Auschwitz, Dachau, Ravensbrück, and Buchenwald, was not a concentration camp. It was an extermination camp. It lay on a desolate, sandy plain 75 miles east of Warsaw, and in the course of thirteen months in 1942–1943, 800,000 people were put to death in the camp's thirteen gas chambers. In his book about Treblinka, Jean-Francois Steiner tells the following story about a Jew by the name of Choken who managed to escape from the camp.

> In the morning he came to Novorgdomsk. He had decided to pass through this city on his way to Warsaw because one of his cousins lived there. He was given a friendly reception, up to the moment he began to talk about Treblinka. The city was languid and peaceful. His cousin begged him not to mention Treblinka to anyone.
>
> "But you believe me, don't you?" said Choken.
>
> "Of course I believe you, but there's no point in frightening the others."
>
> "Is it better to let them die?"
>
> "Not die, hope. Hope is the only thing they have left."
>
> "But wouldn't it be better for them to die like human beings?"
>
> "What difference does it make if you're a human being or a lamb or a Jew, when you have to die in any case?"
>
> The conversation went on in this vein for

several hours, but then Choken went outside and started to speak to the crowd. Fifteen minutes later, the Jewish police arrived and took him away, first to jail and then to the Jewish Council. The chairman of the Council accused him of intentionally creating panic in order to buy gold and jewels at bargain prices. Choken smiled without answering, and on the way back to jail he escaped.

At the beginning of January he arrived in Warsaw. There the Jews listened, often with indifference, sometimes with hostility, but in any case he was allowed to speak.

Politicians Are Also Men—and Act Like Ostriches

Scientists are speaking out more and more often in the general debate on international relations. They produce cold figures and convincing arguments about mankind's vital problems—new data on the effectiveness of the atomic bomb and on the enormous potential of biological warfare as a means of extermination, frightening new predictions about population growth and the food supply crisis. The individual scientist cries out a few times and is quickly used up. The politicians defend themselves in various ways.

"Of course I believe you, but there's no point in frightening the others."—"Is it better to let them die?"—"Not die, hope. Hope is the only thing they have left."—"But wouldn't it be better for them to die like human beings?"—"What difference does it make if you're a human being or a lamb or a Jew when you have to die in any case?"

Or: "You're intentionally creating panic so you can buy up gold and jewels at bargain prices." Read: "You scientists exaggerate and put on airs in order to create a sensation." The politicians must be made to understand that they are responsible for the continued survival of our civilization, and that they are to blame for the fact that the strong nations cynically continue to oppress the weak and force them to live in want. It is cowardly and incompetent for politicians to dismiss constructive proposals for a new world order as unrealistic in reference to the political situation in one particular country or another. Surely politicians are not simply to put up with developments based on a public opinion that is dangerous or stupid. They must realize that one of their tasks is to try to alter public opinion and create new and more appropriate values. They must at least want to do this; otherwise they are of very little use to us.

That politicians lack initiative, or set their goals way too low, probably has something to do with their built-in capacity to reject unpleasantness, a capacity they share with all other human beings including the Jews in Novorgdomsk. This ability of ours to screen ourselves off is deeply rooted and seems to be difficult to come to terms with. However, it is important for politicians to be aware of its existence and its effects. For their part, scientists must do their best to make it hard for politicians to ignore unpleasant facts.

Blind Faith in Progress

If a person turns his gaze away from some particular object, it quickly fastens onto another. This is also true of thoughts. If we want to get away from the unpleasantness in our surroundings, we think about something else. If we do not believe that human civilization is going to perish, then we believe that scientific, technical, and economic progress will come to our rescue, that all "progress" is good. A Norwegian writer named Finn Carling takes roughly the following view.

Anyone who ventures to express doubt as to the blessings of progress can count on being called conservative, reactionary, and old-fashioned. In many cases he probably is, too, but it may also be that blind faith in progress is the result of a deep, unconscious fear of change. Since change cannot be avoided, we assure ourselves and others that all change is absolutely for the best. We avoid a sober evaluation of the consequences of progress, we avoid taking a stand and accepting or rejecting the different steps along the way. The truth is that the catastrophic threats we face today can be blamed on a blind faith in the blessings of progress. Whether it is a question of atomic weapons, pollution, or population, an essential part of the problem is to be found in the various "improvements" man has made. We have been borne along by the faith of several generations in technical progress, and we have been unable or unwilling to see the negative aspects of our interference with nature. The tragedy, of course, is not in progress itself but in the fact that we have been incapable of foreseeing its results. We are all to blame in this, but a particular responsibility rests on the holy men of progress, the scientists and inventors, who have seldom been willing, even when it was possible, to view their own contributions to progress in a larger, objective perspective.

It Is Not Always the Strongest Who Survive—but the Most Well-Adapted

Mankind's ostrichlike mentality makes life hard for modern prophets of doom. If people would open one eye for a moment, they would discover that from one-half to two-thirds of the world's population is already in great distress, diseased and hungry. If we do not need to make a 100 percent investment for our own sake, we ought to be able to make it for the sake of others. Most people in the developed world have an unbelievable double standard in this respect. They simply evade the issue without realizing how cynical such evasion really is. Deep down in these prosperous individuals there may be some terrible, dark thought, some kind of hope that the strongest will survive—although the evaders, so far clearly in the majority, do not want to admit that such a thought exists.

It is quite incorrect to believe that it is always the strongest who leave the field with the biological victory, that is, survival. Survival of the fittest does not mean the survival of the strongest, it means survival of whoever is best adapted to a given situation. And as we stated earlier, surivival is primarily a question of the group—not the individual.

The survival of the strongest seems to have applied to the white race during the historically unique period in which it spread its temporarily superior civilization over the world, annihilating other peoples and cultures. But this is because the white race was the most well-adapted to an era in which there was still unused land to be won. But who knows what will happen next? We in the wealthy nations are singularly poorly adapted to what our civilization has brought about. It may very well happen that what was once an advantage will suddenly become a disadvantage. Perhaps the only survivors of a nuclear war will be the remnants of some tribe driven deep into the jungle. Or maybe some Eskimos up near the North Pole will be the only ones not to burn up. In that case they will survive, not because they were stronger than we, but because they were better adapted to their culture than we are to ours, and because they happened to be out of range of what we brought about.

The Dilemma of Moral Philosophy

Any effort to solve the serious problems we face must begin with a realistic understanding of man himself, of his possibilities and his limitations. Let me pursue my analysis a little further.

At an international seminar on "Ethics and Modern Man" held in Oslo in the spring of 1967, Oxford professor Philip Hallie attacked contemporary moral philosophy on the grounds that it deals with ethical problems in an impersonal, logical, and theoretical form. Moral choices never look that way to the person who is in the midst of an ethical conflict and must decide on a course of action. Having to make a choice is a concrete experience, often difficult and painful, and always highly personal. According to Hallie, moral philosophy depersonalizes an area that is one of the most personal in life, and therefore there is something fundamentally wrong with moral philosophy.

But this failing is built into its assumptions and is probably very hard to get away from—the moral philosopher is not himself involved in the emotional problem he is trying to evaluate. If he were, he would not be able to treat the problem with scientific rigor. Let me develop that idea somewhat by introducing two new concepts: guilt ethics and consequence ethics.

Guilt Ethics and Consequence Ethics

If someone close to me does something that conflicts with my notion of what is right, and if his act affects me personally, then I become angry with that person. I judge his actions in accordance with what I consider to be the degree of his guilt. In other words, my judgment is guilt-ethical. But I judge a criminal, whom I do not know and who sits locked up at a safe distance, in a different way. I may insist that society is treating this man inadequately when I consider the "consequences" of the treatment. At least in part, his punishment is meted out according to guilt-ethical norms. I maintain that this criminal ought to be dealt with according to consequence-ethical norms. The reason that I can take a consequence-ethical view of the criminal but am unable to take anything but a guilt-ethical view of the man who is close to me is simple—I have the criminal at a distance in various ways.

The moral philosopher stands in relation to his scientific problem as the consequence-ethical observer does to the criminal. If the philosopher gets closer to his problem, he no longer functions as a philosopher.

What applies to the individual applies also to a group or a people. It is easier for Swedes to take a consequence-ethical view of the so-called Negro riots in the United States than it is for white Americans. To us, it seems indefensible that American society should attempt to preserve social injustice with weapons. Perhaps most of us can see that the best way of putting an end to black rioting is to give blacks what they want, and a little more. It is much more difficult for the average American to see black protest in this light and to come to a similar conclusion. In rioting, the blacks attack a part of the American tradition. The whites assign them guilt and prepare to defend their property with weapons. Law and order. Despite the fact that this is the worst solution from every point of view. The Swedish view of "looting and burning" is not so different from that of white America. The difference in our judgment of the concrete situation is determined by the fact that we have the problem at a certain distance.

The Place of the Large Brain in Politics

From the evolutionary point of view, our large brain is a supervisory organ of recent origin. It appears that we can put our cerebral capacity to better use when dealing with problems from which we have an emotional distance than when the problems are close at hand. Ultimately, this means that we must give the cerebral functions a greater place in politics, and guard against limiting political activity to that part of the brain that we have in common with the lower animals. We can no longer survive without some control over this limitation in our nature. By artificial means, that is, politically, we must attempt to cope with the negative aspects of our capacity for getting involved, and still try to hold on to its advantages.

The Limitation of the Specialist

Many people have described the shortsightedness and limitations of the specialist. In their book, *M70*, Hannes and Kerstin Alfvén repeat the old adage that war is far too important to be left to the generals. First of all, the general has to hold his own against his colleagues. Secondly, he has to assert the interests of the army, especially those of the officers and the generals, because if these are not provided for, his assignment is impossible. The country's interests finish third. And we dare not even mention the interests of mankind. One would like to see the priorities reversed.

The Alfvéns apply the same reasoning to financiers. Every financier must first of all hold his own against his colleagues, since they could easily ruin him. In the second place, he has to think about working together with his colleagues to protect their common interests against governmental interference, wage demands from employees, and so on. Only in the third place can he think about the general welfare. One would like to see the priorities reversed.

Exactly the same reasoning can be used about research scientists. And sadly but undeniably, it can also be applied to politicians. For the individual politician, the most important problem must be to maintain himself and his party in power. Secondly, he has to assert his country's interests against those of other countries and see to it that his people get their place in the sun. Only in third place can he think about mankind and its future. One would like to see the priorities reversed.

As far as the generals are concerned, the problem has been solved in most countries by putting the politicians in control of the army. In many places, politicians have also had the sense to impose regulations on the world of business. And scientists are given badly needed help in setting the priorities for their research. The big, tough problem is how to put sufficiently effective controls on the politicians, who really wield the power. According to the principles of democracy, it is the people, in elections, who shall exercise control over the politicians. Of course, to a certain extent, this actually happens in many democracies. But in general, the principles are all too easy to get around. Those in power can control the will of the people in many ways. The direct representatives of the people, the members of parliament, are themselves politicians, and moreover, can easily be bypassed by an administration that has the help of experts. Members of parliaments all over Europe complain of such encroachments.

A free press can have a very important controlling function. But the press is all too often a disappointment—the editors become too involved in party tactics to maintain their distance. Quite as much as the politicians themselves, they fall victim to a tactical inability to act. The editorial pages are particularly unreadable just before elections.

The difficulty is to convince politicians that they actually need the help of people partially outside the realm of party politics, and to get them to open themselves voluntarily to criticism from pressure groups outside their own circles to a much greater extent than is now the case. We must create a greater number of pressure groups that lack the decision-making function but that have considerable opportunity to publicize their views in the mass media. These groups must also have a high status, sanctioned by the politicians. It is natural to think first of scientists, who know something about the dynamic forces they are setting free.

Science and the Goals of Society

Johan Galtung, who opened the ethics seminar in Oslo, maintained that social scientists cannot help us formulate society's goals but only indicate the means of achieving the goals society has decided to establish. He feels that the attempt to evaluate various social measures scientifically is usually contrary to the very nature of social choices, which usually rest on nonscientific value judgments. According to Galtung, the development of sociology and social psychology should leave more and more room for nonscientific ethical choices, rather than the other way around, as many have hoped.

I believe Galtung is wrong here. Science has a lot to offer in the way of ethical assessments. I think that a comparison between cultural and biological evolution permits us to draw certain conclusions about the values we ought to seek. Such a comparison lies very much within the domain of science, and specifically within the domain of the social sciences.

Biologically, man is equipped for the life he led before the invention of writing. In creating values for the society of the future, we must begin with the needs and the joys that dominated the life of Stone Age man. This statement already contains a value judgment ultimately bound up with the conviction that there is some worth in man's survival as a species. But we can all presumably agree on that. (We must quite crudely turn a deaf ear to anyone who puts ideology above the survival of mankind.)

It is from this starting point that we should evaluate our technical "progress." In prehistoric society, there was very little specialization—almost everyone understood everyone else's work. Perhaps we should limit specialization in industrialized society. In prehistoric society there were no big cities—maybe we simply can't handle urbanization. The considerations of profitability that seem completely to govern the course of progress in the wealthy nations were utterly foreign to prehistoric man. It seems to me that the profit motive in its modern form is ready for the scrap heap.

The Role of the Scientist in Politics

I realize that politicians and political scientists are irritated by research scientists wanting to take part in the political game like clumsy, well-meaning amateurs. If scientists think politics is so important, say the politicians, then they ought to become politicians and follow the rules. If scientists had to take the political responsibility, they would immediately become more cautious —as cautious as the politicians are forced to be.

The trouble, however, is that as soon as a scientist becomes a politician, he becomes incapable of carrying on the dialogue that must exist between scientists and politicians. He is stricken by all the tactical paralysis of a politician.

Developments have shown quite clearly that, so far, the dialogue between scientists and politicians has not been conducted as publicly as we would like. In the forties, for example, many scientists in Sweden were shouting separately about water pollution. But their voices were all too easily silenced. The problem is to give scientists an important place in politics, yet stay within the framework of democracy and within the framework of the principle that all men have equal worth.

We have touched on the shortsightedness of politicians. The shortsightedness of scientists was only mentioned in passing. Nevertheless, it is a very important concern. The scales of values that obtain in the relatively closed world of scientific research do not agree at all with the scales of values that exist out in society. The mutual adjustment of these scales of values in a public dialogue could be of the greatest possible importance.

Naturally, the content of such a scientific-political dialogue would depend entirely on how gravely we viewed mankind's problems. In my opinion, the dialogue being conducted today on this subject is often much too nationalistic in character. If we are prepared to accept the description of our situation given in this book as more or less accurate, in other words, if we pull our ostrich heads up out of the sand, then we must come to the conclusion that the combined intelligence of the entire world must be applied to these problems. Only an all-out effort will be adequate. We need to mobilize scientists for the common goal in the same way we do at the height of a war. We must establish interdisciplinary teams of natural scientists, sociologists, doctors, engineers, teachers, and lawyers, who can produce new ideas out of our supply of knowledge, create better technical methods, better organizational forms, and work out social innovations that can be applied quickly and broadly enough to be effective. There is no guarantee that it would work, but we do have to make the attempt.

Obviously, scientists alone are not enough. Everyone has to help, especially politicians, but also businessmen, union leaders, workers, and administrators. Scientific efforts will be of no use unless they are supported by assistance and advice from every other group. The rest of society must give scientists a chance to work on the problem in a new and intensive way—as during a war.

There is a widespread and even justified distrust of science and its representatives. After all, it is modern science and technology that have made our problems what they are. But in spite of the fact that it is scientific progress that has created the problems, it may still be that more science is needed to help us past them. The cure for water polluted by detergents might be the invention of better detergents.

In many important areas, the basic knowledge necessary for a rational solution of some problem is in the hands of very few people who are not scientists themselves. New ideas from cell biology, ecology, or ethology could be of decisive importance for solving our biological problems. As for the problems of social organization, it may be that new ideas from the behavioral sciences or from game theory offer the only solutions. We cannot escape the fact that scientists are indispensable.

Science for Survival

Most scientists plan far ahead. From our point of view, much too far ahead. They hope that their careful studies will fit into some large intellectual synthesis sometime in the unspecified distant future. We ought to insist on more, insist on applicable results within the next five years—or why not one year, the way we do in wartime? During the last world war no one had thirty years in which to come up with answers. But the scientists found time for the experiments and new devices that led to radar, sonar, and the atomic bomb in the course of one to four years. We need the same massive mobilization of research today.

In order to achieve this we need to establish stiff priorities. We need to talk about which problems are the most pressing, and which require the greatest investments. At the highest level of urgency is some kind of feedback organization to preserve peace, something sufficiently stabilizing to prevent the big blast within the next five years. Secondly, we need a series of measures to reduce a number of other risks: destruction of the environment, the downfall of democracy, famine, and so forth. We may have differing views as to which degree of urgency to assign the different problems science is currently working on, but the important thing is that an intense discussion of priorities should get under way.

A lot of scientists are more pessimistic than they need to be in the face of these prospects, because they do not understand how quickly social innovations can achieve widespread effects in society. We can name examples like public opinion surveys, operations analysis, information theory, programmed learning. Many such developments have acquired great significance in a short time. And there are other examples, technical inventions with great social impact: computers, automation, the Pill, TV. Most of these became widespread within a period of ten years. The cliché about social lag is no longer true.

Up to now, we have used science for intellectual enjoyment, for controlling nature, for war. Now we must use science to survive.

Scientists today are no more willing to set these kinds of priorities than most other people. In order to live and be happy, we all have to behave like ostriches or else be cynics with a double standard. And that includes you, and me. It is important to remember that we have no completely objective scientists at our disposal. Scientists, like other people, give way to political feelings. I think we can improve the emotional atmosphere between politicians and scientists if we acknowledge this simple fact.

Back to the Whales

Finally, we will return to the whales, and introduce a man who is optimistic about the future and does not hesitate to propose the blue whale (with a brain weighing 13 pounds) as the beef cow of the sea. The optimist in question is named Gifford Pinchot and is a Career Investigator, now looking for a career in whales.

Pinchot begins by pointing out the world's shortage of protein. About one-third of the world's population now suffers from some type of protein deficiency. Half of all the mortality in children between weaning and five years of age depends at least in part on a lack of animal protein in their diets. Pinchot wants to attack this problem by finding more protein for the world's growing population. All life, with some few exceptions, is in the final analysis dependent on solar energy, which transforms inorganic nutritive matter into organic matter by means of photosynthesis. This process is accomplished by plants, both on land and in the sea. If solar energy were not entrapped in this way, the large animal populations would soon die out from lack of food. After a year or so, the last large predators would hold one final but short feast.

Since the oceans cover almost three-fourths of the surface of the earth, they receive three times as much sunlight as the land and therefore have the potential to produce more food than the land. This potential still lies idle—we have not yet learned to sow, fertilize, and harvest in the sea with the same efficiency we can achieve on land. In the sea, we are still hunters, not farmers. Large areas of the sea produce very small amounts of food because the concentration of inorganic nutritive matter is far too low near the surface. Deep water is rich in such nutritive material, but no photosynthesis can take place at such depths because light never penetrates that far.

In the Antarctic, the production of organic matter is very high because of the upsurge of bottom water that is rich in nutrients. There are also cases of the flow of nutrient-rich deep-sea water increasing surface productivity in warmer seas. The Peru Current is perhaps the best example. It comes up along the west coast of South America and in the vicinity of the Galapagos Islands. The productivity of these waters is very high. Tuna fishing in the vicinity of the Galapagos Islands is so profitable that it attracts fishermen from great distances, and the anchovy fisheries off the coast of Peru are among the best fisheries in the world.

Pinchot realizes that direct enrichment of tropical seas would be too expensive and inefficient. Nor would it pay simply to pump up water from the bottom; it is colder than surface water and would only sink again. (Though he does not indicate the slightest hesitation from an ecological point of view.)

The solution, according to Pinchot, is to be found in the coral atolls. In an atoll, the coral reef surrounds the central lagoon like a fence. There are a great many atolls in the Pacific Ocean, varying in size from a couple of thousand yards in diameter to 460 square miles. The depth of the lagoon is in proportion to the size, the average depth being somewhere around 65 feet. This kind of lagoon makes an ideal petri dish for water-enrichment experiments; the wind even helps by stirring.

Pinchot comes to the conclusion that the best means of enrichment is to pump in deep-sea water from the ocean outside. The pumps can be driven by the prevailing winds. Just because a man believes wholeheartedly in progress doesn't mean he lacks a creative imagination!

The Blue Whale as the Beef Cow of the Sea

Phytoplankton will now grow in the fertilized lagoon. This in turn will be eaten by zooplankton, and then the plantation will be completed by the addition of some animal that refines protein and can be harvested. After diverse deliberations, which appear to be sound, Pinchot comes to the conclusion that baleen whales are the best animals to harvest. They live directly on zooplankton and consequently there is no need for uneconomical middlemen. They are also so large that they are relatively resistant to predators. The blue whale, the largest of all whales, is a particularly good choice from another point of view, says Pinchot: since they have been hunted and are hunted so intensively, there is a risk that they will die out. Tragically enough, Pinchot is probably right in thinking that there are no other ways of preventing this. Experience shows us that whalers will not give up until it no longer pays to capture whales. An expert has recently asserted that there are now so few blue whales that only drastic measures can save the whaling industry from complete collapse. As if it were the whaling industry that mattered!

Immediately after remarking on mankind's shortsightedness, Dr. Pinchot goes on to say that these filter whales constitute a unique link in the oceanic food chain in that they eat zooplankton. If they are exterminated, this extremely effective mechanism for the transformation of plants into animal protein will be lost forever. Moreover, baleen whales can transform plankton into fat and many other biologically (?) useful products, such as hormones (read: into many other products of apparent short-term usefulness to man). Pinchot continues: The extermination of the whales would be really tragic for mankind at a time when the number of people on earth is increasing explosively, and we are threatened by a shortage of protein. This could be prevented by raising whales in coral atolls in the same way the American buffalo was saved by breeding it in captivity.

This all sounds plausible. But it is infinitely more important that we use our strength to prevent population increase by setting a good example in the wealthy nations, where the population density is very high. Also, it is more important to stop killing excessive numbers of whales than to raise them in captivity. Dr. Pinchot will perhaps defend himself by saying that one good deed does not prevent another. That may be true, but it is so easy for us to forget the other part, the hard part. It is to talk about the hard part that this book is being written.

Dr. Pinchot concludes by pointing out that it would be enormously advantageous for scientists to be able to study these, the largest of all animals, in a way they could never be studied in the wild state. Blue whales might prove to be more interesting than dolphins from the point of view of the behavioral sciences. On this point I am in complete agreement with our friend, the believer in progress, but it is tragic that we have to view this as a by-product to the use of blue whales as the beef cows of the sea. I believe our experiments would yield better results if we approached the whale with greater respect.

Man's Place on Earth Once More

I am now back where this book began. We must start thinking along new lines about our place on earth. In the first place, we do not own it. We are not its masters. It was not created for us to play with, or consume, or vandalize. We are a part of it. We came out of it, and we ought to be proud and humble as we trace ourselves back to the first complex molecules in the first seas. The earth evolves, and in this stage of evolution, *Homo sapiens* has come to dominate almost all life, but we ourselves are caught up in the earth's evolutionary process.

If we could learn to see our place on earth this way, we wouldn't dream of turning the blue whale into a beef cow. And we would approach the sperm whale with a new respect, which would very likely bear rich fruit.

After reading my denunciation of Pinchot, one of my more realistic friends said, "You know, what you've done is to conjure up a new sacred cow alongside the one in India—the sacred whale cow." This puts its finger on the double standard we all embody, and which to some extent characterizes even this book. No matter how much we try to comprehend our place in the biological scheme, and inspire respect in ourselves for other forms of life, we are doing it, deep down, for our own sakes, and on the basis of an anthropocentric picture of the world. There is no denying it. Let us admit that all notions about sacred cows include an element of the double standard. It is hard enough for us human beings, both singly and in groups, to feel respect for *each other's* individuality. But let us hope that biological insight will help us in some small way to acquire this kind of respect. Biological insight might also lead us to reduce the pace at which we are sawing away at the limb we're sitting on.

Even people who understand the doctrine of evolution intellectually have a hard time *feeling* strongly enough that man is a part of the living earth. Man needs something to shake him to his roots, to impress him deeply with the fact that he does not own the earth. This something, which might give man the humility he needs, could be a greater knowledge of the other large-brained animals.

Bibliography

Alpers, Antony. *Dolphins: The Myth and the Mammal.* Boston: Houghton Mifflin Co., 1961.

Andersen, Harald T. *The Biology of Marine Mammals.* New York: Academic Press, 1968.

Ardrey, Robert. *The Territorial Imperative: A Personal Inquiry into the Animal Origins of Property and Nations.* London: Collins, 1966.

Bateson, Gregory. "Problems in Cetacean and Other Mammalian Communication." In Kenneth S. Norris, ed., *Whales, Dolphins, and Porpoises.* Berkeley: University of California Press, 1966. Pp. 569–79.

Bonner, John T. *The Ideas of Biology.* Modern Science Series. New York: Harper & Row, Publishers, 1961.

Breland, K. *Animal Behavior.* New York: Macmillan Co., 1966.

Chance, M. R. A., and A. P. Mead. "Social Behavior and Primate Evolution." In *Evolution,* Symposia for the Society of Experimental Biology, Vol. 7. London: Academic Press, 1953.

Chapman, Seville. "Dolphins and Multifrequency, Multiangular Images." *Science* 160(1968): 208–9.

Devine, Eleanore, and Martha Clark. *The Dolphin Smile.* New York: Macmillan Co., 1967.

Fog, Mogens. *Hvordan hjernen arbejder* [How the Brain Works]. Copenhagen: Munksgaards Forlag, n.d.; Stockholm: Wahlström & Widstrand, 1966.

Gardner, R. Allen and Beatrice T. "Teaching Sign Language to a Chimpanzee." *Science* 165(1969): 664–72.

Howell, F. Clark. *Early Man.* Life Nature Library Series. New York: Time-Life Books, 1965.

Kellogg, Winthrop N. "Communication and Language in the Home-raised Chimpanzee." *Science* 162(1968): 423–27.

Koestler, Arthur. *The Ghost in the Machine.* London: Hutchinson, 1968.

Lenneberg, Eric H. *Biological Foundations of Language.* New York: John Wiley & Sons, 1967.

Lilly, John C. *The Mind of the Dolphin.* New York: Doubleday & Co., 1967.

Lorenz, Konrad. *On Aggression.* London: Methuen, 1966.

Melville, Herman. *Moby Dick.* London: Pan Books. (Originally published 1851.)

Morris, Desmond. *The Human Zoo.* London: Cape, 1969.

Norris, Kenneth S., ed. *Whales, Dolphins, and Porpoises.* Berkeley: University of California Press, 1966.

Pinchot, Gifford B. "Whale Culture." *Perspectives in Biology and Medicine* (1966): 33.

Platt, John. "What We Must Do." *Science* 166 (1969): 1115–1121.

Psychobiology, the Biological Bases of Behavior: Readings from Scientific American Magazine. San Francisco: W. H. Freeman & Co., Publishers, 1967.

Rabb, George B. "Social Relationships in a Group of Captive Wolves." *American Zoologist* 7 (1967): 305–11.

Savage, J. M. *Evolution.* New York: Holt, Rinehart & Winston, 1963.

Slijper, E. J. *Whales.* New York: Basic Books, 1962.

Thomas, Lewis *Perspectives in Human Biology.* Gainesville Address, 1967.

Tiger, Lionel. *Men in Groups.* New York: Random House, 1969.
Williams, J. H. *Elephant Bill.* London: Hart-Davis, 1955.
Wooldridge, Dean E. *The Machinery of the Brain.* New York: McGraw Hill Book Co., 1963.